AF412085

Contemporary Clinical Neuroscience

More information about this series at http://www.springer.com/series/7678

Noriyuki Koibuchi • Paul M. Yen

Editors

Thyroid Hormone Disruption and Neurodevelopment

 Springer

Editors
Noriyuki Koibuchi
Department of Integrative Physiology
Gunma University Graduate School of
  Medicine
Maebashi, Gunma, Japan

Paul M. Yen
Laboratory of Hormonal Regulation
Cardiovascular and Metabolic Disorders
  Program
Duke-NUS Graduate Medical School
Singapore, Singapore

Contemporary Clinical Neuroscience
ISBN 978-1-4939-3735-6   ISBN 978-1-4939-3737-0   (eBook)
DOI 10.1007/978-1-4939-3737-0

Library of Congress Control Number: 2016940093

This Springer imprint is published by Springer Nature
The registered company is Springer Science+Business Media LLC
The registered company address is: 233 Spring Street, New York, NY 10013, U.S.A.

Contents

Contributors

Takaaki Abe Division of Nephrology, Endocrinology, and Vascular Medicine, Department of Medicine, Tohoku University Graduate School of Medicine, Sendai, Miyagi, Japan

Division of Medical Science, Tohoku University Graduate School of Biomedical Engineering, Sendai, Miyagi, Japan

Department of Clinical Biology and Hormonal Regulation, Tohoku University Graduate School of Medicine, Sendai, Miyagi, Japan

Paolo Beck-Peccoz Department of Clinical Sciences and Community Health, University of Milan, Fondazione IRCCS Cà Granda Ospedale Maggiore, Milan, Italy

M. Bialy Department of Experimental and Clinical Physiology, Laboratory of Center for Preclinical Research, Medical University of Warsaw, Warsaw, Poland

Gregory A. Brent Department of Medicine, David Geffen School of Medicine at UCLA, Los Angeles, CA, USA

Irene Campi Department of Clinical Sciences and Community Health, University of Milan, Fondazione IRCCS Cà Granda Ospedale Maggiore, Milan, Italy

F. Chatonnet Institut de Génomique Fonctionnelle de Lyon, Université de Lyon, Université Lyon, INRA, CNRS, Ecole Normale Supérieure de Lyon, Lyon, Cedex, France

Ines Donangelo Endocrinology Division, Allegheny General Hospital, Pittsburgh, PA, USA

F. Flamant Institut de Génomique Fonctionnelle de Lyon, Université de Lyon, Université Lyon, INRA, CNRS, Ecole Normale Supérieure de Lyon, Lyon, Cedex, France

Kingsley Ibhazehiebo Department of Integrative Physiology, Gunma University Graduate School of Medicine, Maebashi, Gunma, Japan

Sumiyasu Ishii Department of Medicine and Molecular Science, Gunma University Graduate School of Medicine, Maebashi, Gunma, Japan

Toshiharu Iwasaki Department of Integrative Physiology, Gunma University Graduate School of Medicine, Maebashi, Gunma, Japan

Noriyuki Koibuchi Department of Integrative Physiology, Gunma University Graduate School of Medicine, Maebashi, Gunma, Japan

Jens Mittag Center of Brain Behavior and Metabolism, University of Lübeck, Lübeck, Germany

Masami Murakami Department of Clinical Laboratory Medicine, Gunma University Graduate School of Medicine, Maebashi, Japan

S. Richard Institut de Génomique Fonctionnelle de Lyon, Université de Lyon, Université Lyon, INRA, CNRS, Ecole Normale Supérieure de Lyon, Lyon, Cedex, France

E.M. Sajdel-Sulkowska Department of Psychiatry, Harvard Medical School and Brigham and Women's Hospital, Boston, MA, USA

Department of Experimental and Clinical Physiology, Laboratory of Center for Preclinical Research, Medical University of Warsaw, Warsaw, Poland

Department of Physiological Sciences, Warsaw University of Life Sciences, Warsaw, Poland

David S. Sharlin Department of Biological Sciences, Minnesota State University, Mankato, MN, USA

Noriaki Shimokawa Department of Integrative physiology, Gunma University Graduate School of Medicine, Maebashi, Gunma, Japan

Rohit Anthony Sinha Laboratory of Hormonal Regulation, Cardiovascular and Metabolic Disorders Program, Duke-NUS Graduate Medical School, Singapore, Singapore

Takehiro Suzuki Division of Nephrology, Endocrinology, and Vascular Medicine, Department of Medicine, Tohoku University Graduate School of Medicine, Sendai, Miyagi, Japan

Department of Medical Science, Tohoku University Graduate School of Biomedical Engineering, Sendai, Miyagi, Japan

Masanobu Yamada Department of Medicine and Molecular Science, Gunma University Graduate School of Medicine, Maebashi, Gunma, Japan

Paul M. Yen Laboratory of Hormonal Regulation, Cardiovascular and Metabolic Disorders Program, Duke-NUS Graduate Medical School, Singapore, Singapore

R. Zabielski Department of Physiological Sciences, Warsaw University of Life Sciences, Warsaw, Poland

Chapter 1
Mechanisms for Thyroid Hormone Action in the CNS

Rohit Anthony Sinha and Paul M. Yen

Abstract This chapter aims to summarize our recent understanding about the action of thyroid hormone (TH) in the CNS. The topics will include the molecular action of TH at genomic and non-genomic levels and its impact on the physiology of brain.

Keywords Thyroid hormone (TH) • Thyroid hormone receptors (TR) • Brain • Neurons • Epigenetic • Deiodinases

1.1 Introduction

Thyroid hormones (THs) mediate important physiological processes such as development, growth, and metabolism in virtually all tissues of the body (Brent 2012; Cheng et al. 2010; Oetting and Yen 2007). There are two major THs secreted by the thyroid gland, levothyroxine (T_4) and triiodothyronine (T_3), with the latter serving as the more biologically active form. Their serum concentrations are tightly regulated by the hypothalamic/pituitary/thyroid (HPT) axis. THs are transported by specific proteins via the circulation to tissues throughout the body and also must pass the blood/brain barrier for delivery to the CNS. Intracellular uptake of TH occurs by specific TH transporters, and the intracellular concentration of TH is further regulated by intracellular deiodinases that convert T_4 to T_3 to increase the TH activity or transform the THs to inert metabolites to reduce it. Intracellular TH then binds to nuclear thyroid hormone receptors (TRs), members of the nuclear receptor superfamily, that are ligand-dependent transcription factors to regulate positive or negative transcription of target genes. There are two major isoforms, TRα and TRβ, which each has different tissue distributions. TRs typically form heterodimers with another member of the nuclear receptor superfamily, RXR, and bind to specific

R.A. Sinha, Ph.D. (✉) • P.M. Yen, M.D.
Laboratory of Hormonal Regulation, Cardiovascular and Metabolic Disorders Program,
Duke-NUS Graduate Medical School, 8 College Road, Singapore 169857, Singapore
e-mail: Rohit.sinha@duke-nus.edu.sg; anthony.rohit@gmail.com

© Springer Science+Business Media New York 2016
N. Koibuchi, P.M. Yen (eds.), *Thyroid Hormone Disruption and Neurodevelopment*,
Contemporary Clinical Neuroscience, DOI 10.1007/978-1-4939-3737-0_1

DNA sequences, thyroid hormone response elements (TREs), that commonly are located in the promoter regions of target genes. TRs activate transcription by recruiting coactivator complexes that contain histone acetyltransferase and methyltransferase activities to alter regional chromatin structure and enable recruitment of general transcription factors to the transcriptional start site of the target gene promoter. Of note, TRs also can repress transcription of target genes in the absence of TH by recruiting corepressors and histone deacetylases. Additionally, TRs also can mediate TH-dependent negative regulation of transcription, although the precise mechanism is not well understood.

Besides binding to nuclear TRs, there also is emerging evidence that THs or their metabolites can bind to cytosolic TRs or other membrane-bound or intracellular proteins to mediate cellular actions. Therefore, given the complexity of the physiological regulation of THs, their delivery and intracellular uptake/concentration to various tissues, and the multiple proteins involved in their transcriptional regulation of target genes, it is not surprising that endocrine disruptors and drugs can interfere with TH action in the brain as well as in other tissues at many points from the central hypothalamic regulation of thyrotropin-releasing hormone (TRH) to target gene transcription. This chapter will provide an introductory background to some of the critical checkpoints that regulate TH action.

1.2 The Hypothalamic/Pituitary/Thyroid Axis

The median eminence of the hypothalamus secretes the tripeptide, TRH, which is transported via the hypothalamic-hypophyseal portal system to the pituitary (Fekete and Lechan 2014). TRH binds to the TRH receptor, a G protein-coupled receptor expressed in thyrotrophs located within the anterior pituitary. TRH binding to the receptor causes activation of phospholipase C resulting in hydrolysis of PIP2 to IP3 and generation of diacylglycerol (DAG). The increase in IP3 stimulates release of intracellular Ca^{++}, which in turn induces thyroid-stimulating hormone (TSH) release from pituitary thyrotroph cells. DAG also stimulates PKC, leading to increased synthesis of both TSH subunits. TSH is a heterodimer comprised of two glycoprotein subunits – the glycoprotein hormone α-subunit that also is found in luteinizing-stimulating, follicle-stimulating, and human choriogonadotropic hormones and the TSHβ-subunit that is unique for TSH.

After TSH is released into the circulation, it eventually binds to TSH receptors located primarily in the thyroid gland. Of note, there have been reports of TSH receptors present in other tissues, including the brain, but their significance is not well understood (Bockmann et al. 1997; Smith 2015). The TSH receptor also is a G protein-coupled seven-transmembrane protein and generates cAMP upon ligand binding. Its activation leads to thyrocyte proliferation, thyroglobulin (Tg) and sodium-iodide symporter (NIS) gene transcription, and stimulation of TH synthesis and secretion.

Approximately 93 % of TH secreted by thyroid gland is T_4, whereas only 7 % is T_3. However, the deiodinases, *Dio 1* and *Dio 2*, that are located in the peripheral tissues, are able to convert much of the T_4 to T_3. Circulating T_3 also negatively regulates TRH

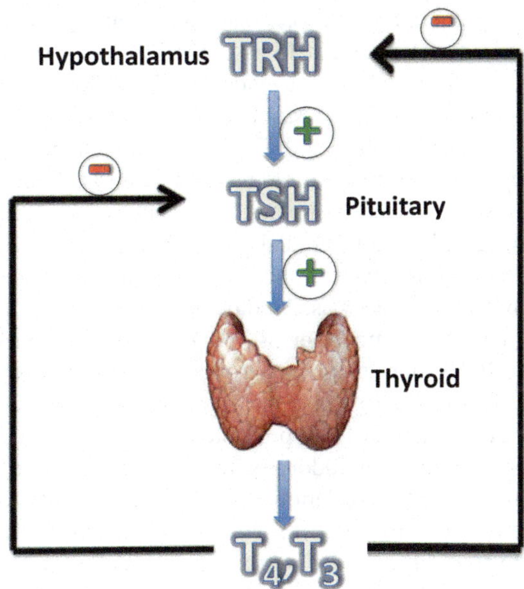

Fig. 1.1 HPT axis. Thyroid homeostasis involves a multi-loop feedback system known as the hypothalamic/pituitary/thyroid (HPT) axis that is found in almost all higher vertebrates. The hypothalamus senses low circulating levels of thyroid hormones (THs; triiodothyronine (T_3) and thyroxine (T_4)) and responds by secreting thyrotropin-releasing hormone (TRH). TRH, in turn, stimulates the pituitary to secrete thyroid-stimulating hormone (TSH). The TSH stimulates the thyroid gland to produce T_3 and T_4 until levels in the blood return to normal. TH exerts negative feedback control over both the hypothalamus and anterior pituitary and thus controls the release of both TRH from the hypothalamus and TSH from the anterior pituitary gland due to negative regulation of these genes by TH-bound TRs

and TSH secretion by the hypothalamus and pituitary, respectively. Thus, the HPT axis is a tightly controlled and coordinated system in which serum TH levels positively regulate TH secretion and serum TH concentration negatively regulates TRH and TSH secretion (Fig. 1.1). A critical stage for this regulation occurs shortly after birth when there is a transient central regulation of the neonatal surge in circulating TH that is critical for normal brain development in mammals. In congenital hypothyroidism, inadequate thyroid hormone secretion due to genetic or developmental defects leads to mental retardation (Garcia et al. 2014).

1.3 TH Synthesis

TH synthesis in the thyroid gland requires several steps that include uptake of iodide by active transport, thyroglobulin (Tg) biosynthesis, oxidation and binding of iodide to thyroglobulin (organification), and oxidative coupling of two iodotyrosines into iodothyronines (Garcia et al. 2014; Grasberger and Refetoff 2011). The thyroid gland is composed of thyroid cells arranged as follicles with their basal surfaces pointing outward toward the capillaries and the apical surfaces pointing inward toward a central lumen containing colloid where nascent TH is generally stored. The synthesis of TH in the thyroid follicular cell is unipolar as evidenced by initial iodide uptake via the sodium-iodide symporter (NIS) in the basolateral plasma membrane and the tyrosyl iodination and iodotyrosine coupling of thyroglobulin at the apical surface of the cell. The iodide uptake by NIS leads to a 30-fold increase in intracellular iodide concentration in thyrocytes vs. serum (iodide trapping). TSH acutely increases the gene expression of NIS; however, iodide overaccumulation within the thyrocyte is prevented via a sensitive negative feedback mechanism in which high intracellular iodide downregulates NIS expression leading to decreased iodide uptake (Wolff-Chaikoff effect).

Thyroglobulin (Tg) is a homodimer of 330 kDa glycoproteins that is secreted into the lumen of follicles, and its tyrosyl groups serve as substrates for iodination and TH formation. Thyroid peroxidase (TPO) is an enzyme located at the apical plasma membrane that reduces H_2O_2 to promote iodide oxidation and binding to Tg as well as iodotyrosine coupling to iodotyrosines. The H_2O_2 within the thyroid gland is generated from NADPH oxidation by the dual oxidases (DUOX) and THOX. Synthesized THs conjugated to Tg are stored within the central colloid lumen formed by thyroid follicular cells to form a reservoir for TH secretion.

Release of TH from the thyroid gland is stimulated by TSH and requires the endocytosis of the Tg-conjugated THs from the colloid lumen. This process is followed by intracellular proteolysis of Tg and hydrolysis of THs from the Tg remnants, eventuating in the release of THs from the thyrocytes. Of note, the polarity of TH secretion is in the opposite direction than TH synthesis. TSH stimulates the proteolysis of Tg and hydrolysis of TH from Tg to enable rapid release of T_3 and T_4 into the circulation. Some Tg also accompanies TH into the serum so the serum Tg level can serve as a useful index for assessing the functional state of the thyroid gland. Genetic defects in any of the key proteins involved in TH synthesis, particularly those for NIS, TPO, Duox, and Tg, can cause goiter and hypothyroidism (Garcia et al. 2014). Propylthiouracil and carbimazole also block iodide organification and iodotyrosine coupling (Reader et al. 1987). Likewise, the foregoing proteins involved in TH uptake, synthesis, and secretion in the thyroid also are potential targets for endocrine disruptors.

1.4　TH Transport in Circulation

Approximately 0.03 % of the total serum T_4 and 0.3 % of the total serum T_3 are present in free or unbound form in man (Schussler 2000). The major serum TH-binding proteins are thyroxine-binding globulin (TBG or thyropexin), transthyretin (TTR or thyroxine-binding prealbumin (TBPA)), and albumin (HSA, human serum albumin). TBG binds 75 % of serum T_4, whereas TTR and HSA bind only 20 % and 5 %, respectively. TH binding to these carrier proteins ensures an even distribution and delivery of hormone throughout the body. The free serum T_3 and T_4 (fT_3, fT_4) levels represent the actual amount of TH available to cells via the circulation and are often used clinically to assess the TH status of patients (Fig. 1.2). TBG production can be

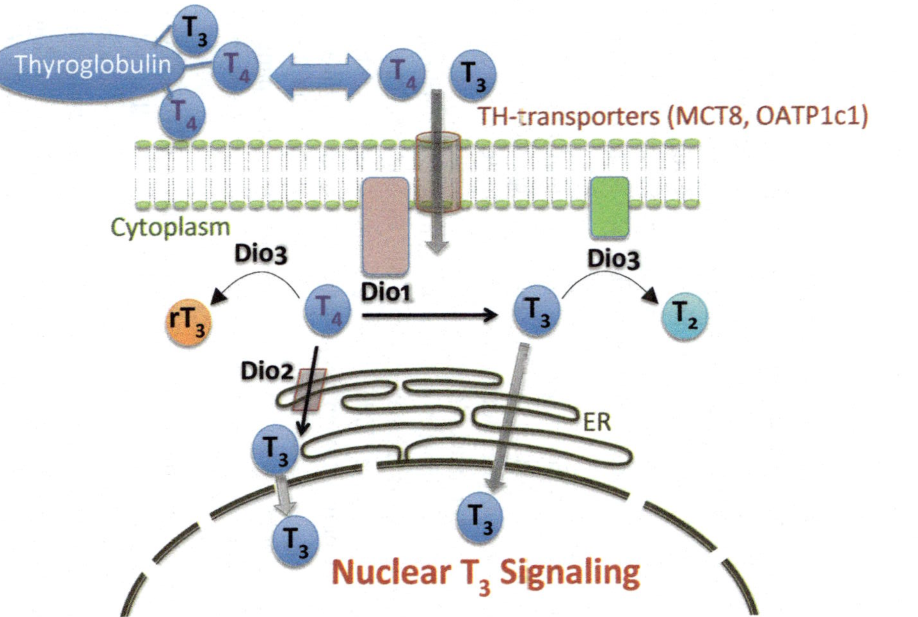

Fig. 1.2 Thyroid hormone transport and metabolism. Most of the thyroid hormone (TH) circulating in the blood is bound to transport proteins such as thyroglobulin, albumin, and transthyretin. Only a very small fraction of the circulating hormone is free (unbound) and biologically active, and there is an equilibrium between bound and free forms of T_3 and T_4 in the serum. TH is transported intracellularly by TH transporters across the plasma membrane. Several transporter families, namely, monocarboxylate transporter (MCT)8, MCT10, and organic anion-transporting polypeptide (OATP)1C1, demonstrate a high degree of specificity toward THs. TH activation and inactivation inside the cells require selenocysteine-dependent iodothyronine deiodinases (DIOs) that catalyze the release of iodine directly from the thyronine hormones. DIO1 and DIO2 catalyze the conversion of T_4 to T_3 and DIO3 inactivates T_4 and T_3. DIO2 is localized to the ER membrane, while DIO1 and DIO3 are found in the plasma membrane

modified by other factors such as estrogen levels, corticosteroid levels, or liver failure. Certain drugs such as aspirin, carbamazepine, furosemide, and phenytoin can displace T_4 from these proteins and raise serum fT_4 levels; however, regulated conversion of T_4 to T_3 generally offers some protection from hyperthyroidism in these cases. Interestingly, polychlorinated biphenyls (PCBs) have been shown to bind to TTR and TBG in vitro (Cheek et al. 1999).

1.5   TH Transporters

Recently, MCT8/10, OATP-1c1, and System L amino acid transporters have been reported to be TH transporters that regulate T_4 and T_3 uptake into cells (Heuer and Visser 2013) (Fig. 1.2). The discovery of specific transporters that are differentially expressed in various tissues raises the possibility that endocrine disruptors may selectively block uptake of TH into certain tissues and decrease intracellular TH concentration. This, in turn, could decrease gene expression of TH target genes in specific tissues.

MCT8 and MCT10 are members of the monocarboxylate transporter family and have 12 putative transmembrane domains with their amino- and carboxy-terminal ends embedded in the plasma membrane. Both transporters promote intracellular uptake of T_4 and T_3 and likely function in a bidirectional manner since they also can facilitate efflux of T_4 and T_3. MCT10 appears to transport T_3 more efficiently than MCT8, whereas MCT8 can transport T_4 better. Both MCT8 and MCT10 are expressed in many tissues such as the liver, kidney, and heart; however, only MCT8 is highly expressed in the human brain. Human mutations in MCT8 have been associated with the congenital syndrome, x-linked mental retardation, and neurologic deterioration (Allan-Herndon-Dudley syndrome) in which affected male patients progressively develop spastic paralysis and cognitive degeneration (Dumitrescu and Refetoff 2013).

Organic anion-transporting polypeptides (OATPs) are members of the SLCO family and can increase intracellular concentrations of amino acids as well as iodothyronines and their sulfate conjugates (Heuer and Visser 2013). The relative distribution of the OATPs varies among different tissues with OATP1B1 and 1B3 expressed specifically in the liver; OATP1A2 expressed in the brain, liver, and kidney; and OATP1C1 expressed only in the brain and testis. Interestingly, OATP1C1 has high specificity and affinity for T_4 and rT_3 and is localized preferentially in the capillaries. Thus, OATP1C1 may be important for transport of T_4 across the blood/brain barrier (Heuer and Visser 2013).

Amino acid transporters that belong to the SLC7 family and the L- and T-type amino acid transporters also can increase TH uptake into cells (Heuer and Visser 2013). L-type transporters promote uptake of large neutral, branched-chain and aromatic amino acids, whereas T-type transporters mediate uptake of the aromatic amino acids Phe, Tyr, and Trp. These amino acid transporters can form heterodimers with another glycoprotein belonging to the SLC3 family, 4F2hc, to form transport channels. The TH specificity of these channels appears to be highest for T_2 and lowest for T_3 and T_4. These channels are widely expressed throughout the body and also are thought to play an important role in the maternal transfer of TH in the placenta during pregnancy.

1.6   TH Deiodinases

The majority of TH secreted by the thyroid gland is T_4 and is frequently considered a pro-hormone for T_3 since it has much lower biological activity and binding affinity for TRs than T_3. Since relatively smaller amounts of T_3 are secreted by the thyroid gland, serum T_3 is mostly derived from T_4 to T_3 conversion by 5′ monodeiodinases (Fig. 1.2). There are three deiodinases (*Dio 1*, *2*, and *3*) that regulate both circulating TH and intracellular T_4 and T_3 levels (Larsen and Zavacki 2012). *Dio 1* is expressed mainly in the liver, kidney, and thyroid and contributes to the peripheral T_3 production and clearance of plasma rT_3. *Dio 2* is primarily expressed in the pituitary, brain, muscle, and brown fat and also contributes to the production of circulating T_3. *Dio 3* degrades TH by converting T_4 and T_3 to T_2, an inert TH metabolite. The brain is the major tissue that expresses *Dio 3* in the body and thus may serve as the main site for clearance of serum T_3 and production of serum rT_3. The placenta also has high levels of *Dio 3* that may protect the fetus from exposure to maternal levels of circulating TH during early development.

The deiodinases also determine the intracellular concentration of T_3 to modulate sensitivity TH within the cell. For example, tissues that express high levels *Dio 2* may respond better to higher circulating T_4 levels than tissues that express low levels due to their ability to convert T_4 to T_3. Thus, while serum TH levels may reflect the integrity of the HPT axis, the actual intracellular TH concentrations that are regulated by deiodinases may be more important in determining TH action within particular tissues. Several drugs such as amiodarone, cholecystographic agents ipodate and iopanoic acid, propylthiouracil, and propranolol are known to block T_4 to T_3 conversion and reduce serum T_3 levels (Burger et al. 1981). Accordingly, endocrine disruptors that block *Dio 1* or *Dio 2* or increase *Dio 3* activities may not only reduce serum T_3 levels but also reduce the effective concentration of T_3 within certain tissues.

1.7   Thyroid Hormone Receptors

TH receptors (TRs) are the cellular homologs of a viral oncogene product associated with chick erythroblastosis, *v-erbA*. TRs belong to the nuclear hormone receptor superfamily that includes the steroid, vitamin D, retinoic acid (RARs), retinoid *X* (RXRs), and liver *X* receptors (Brent 2012; Cheng et al. 2010; Oetting and Yen 2007). They are encoded at two genomic loci (TRα and TRβ) located on human chromosomes 17 and 3 that express two major isoforms, TRα and TRβ. Similar to other nuclear hormone receptors, TRs have a central cysteine-rich DNA-binding domain (DBD) in which two zinc ions are each complexed with four cysteines to form two zinc "fingers" that enable TR binding to DNA. They also have a carboxy-terminal ligand-binding domain (LBD) that binds TH. Additionally, TRs have a nuclear translocation signal sequence located just behind the DBD. Furthermore, TRs have 12 amphipathic helices in the LBD that provide contact surfaces for interactions with RXR as well as the coactivators and corepressors that transcription. The x-ray crystal structure of the LBD has provided considerable insight into the structure of the "ligand-binding pocket" as well as the interaction sequences with RXR and other transcriptional cofactors. In particular, ligand binding to TR induces major conformational changes in the LBD, especially in helix 12, that render corepressor interaction surfaces less favorable for binding to corepressors and position TR coactivator sequences to bind with coactivators (Fig. 1.3).

There are additional isoforms of TRα and TRβ that modulate TH hormone action in different tissues. TRα-2 shares the first 370 amino acid sequence with TRα-1 before the LBD diverges completely due to alternate splicing of exons. Since TRα-2 has the same DBD as TRα but cannot bind TH, it has been suggested that it might be a competitive inhibitor of TRs for binding to TREs and thereby reduce TH action in tissues where it is highly expressed such as the brain and testis. Another nuclear receptor, Rev-erbA, is encoded within the TRα gene but is transcribed from the DNA strand opposite of that used to generate TRα-1 and TRα-2. Rev-*erbA* is highly expressed in adipocytes, muscle, and the liver and may mediate circadian transcriptional TRβ effects in its target genes. TRβ gene encodes two TR isoforms, TRβ and TRβ-2, due to transcription from two different start sites. TRβ is expressed in almost all tissues, whereas TRβ-2 is selectively expressed in the pituitary, retina, and hypothalamus. While the precise role of TRβ-2 is not known, it is possible that it may mediate critical developmental or tissue-specific functions.

The two major TR isoforms, TRα and TRβ, share significant amino acid sequence homology in their respective DBDs and LBDs (Aranda et al. 2013). Moreover, they both have similar binding affinities for TH and appear to regulate overlapping sets of target genes. These features notwithstanding TRβ-selective ligands recently have been developed that can regulate transcriptional activity in tissues where TRβ is selectively expressed (e.g., the liver) with little effects on tissues that express predominantly TRα (e.g., the heart and bone). These isoform- or tissue-selective compounds hold promise for decreasing cholesterol, obesity, and nonalcoholic fatty liver disease. On the other hand, it is possible that there may be isoform- or tissue-selective endocrine disruptors that affect TH functions in some tissues but not others.

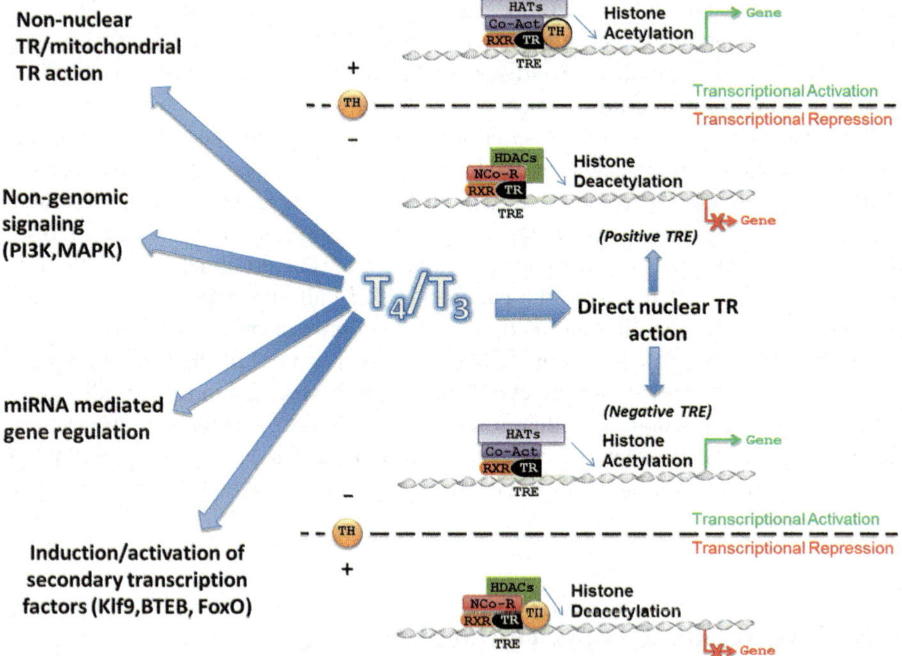

Fig. 1.3 Thyroid hormone signaling. The majority of TH signaling involves binding to nuclear receptors (TRs) that regulate gene expression by binding to thyroid response elements (TREs) in DNA either as monomers, heterodimers with retinoid X receptor (RXR). Gene promoters which are activated by hormone binding to TR are known to harbor positive TREs, whereas those which are repressed in the presence of hormone have negative TREs such as those found on TRH and TSH genes. On positive TREs containing genes in the absence of hormone, TRs recruit corepressor (e.g., NCoR, SMRT) and histone deacetylase (HDACs) proteins to modify chromatin near the promoter into a transcriptionally inactive state. Binding of thyroid hormone results in a conformational change in TR which displaces corepressor from the receptor/DNA complex and recruitment of coactivator proteins and histone acetylase (HATs). The DNA/TR/coactivator complex then recruits RNA polymerase that transcribes downstream DNA into messenger RNA and eventually protein that results in a change in cellular function. The converse is thought to occur on genes with negative TREs, although the precise mechanisms are not well understood. Besides the classical nuclear TR signaling, T_3/T_4 also may induce either activation or expression of other transcription factors (KLF9, FoxO1) that mediate induction of late-responsive target genes. TRs also have been located on extranuclear sites such as mitochondria and have been implicated to regulate direct mitochondrial functions. Additionally, posttranscriptional regulation of gene expression via miRNAs has been shown to mediate TH effects on transcription nonclassical signaling. Additionally, T_3 and T_4 can induce rapid non-genomic signaling in cells which may or may not involve TRs

1.8   TR Binding to Thyroid Hormone Response Elements

TRs are ligand-regulated transcription factors that form heterodimers with another member of the nuclear hormone receptor superfamily, RXR. The TR/RXR heterodimer binds to DNA sequences, thyroid hormone response elements (TREs), both in the absence and presence of TH (Aranda et al. 2013; Brent 2012; Oetting and Yen 2007)

(Fig. 1.3). TREs typically are composed of two hexamer sequence AGGT(C/A)A (half-sites) arranged as direct repeats with a spacing of four nucleotides that induce TR/RXR heterodimer binding polarity, with TR binding to the downstream half-site and RXR binding to the upstream half-site. However, TRs also can bind to other arrangements of half-sites such as palindromes and inverted repeats as well as to half-sites that contain significant degeneracy from the canonical hexamer sequence. Recent chIP-seq analyses of TR binding throughout the whole genome have demonstrated that TRs can bind to DNA sequences not only in the promoter region but also to far upstream, intronic, and 3′ sequences of target genes. Additionally, TRs bind to sequences that do not resemble the canonical TRE half-site sequences, raising the possibility that TRs can bind indirectly to DNA via protein-protein interactions with other DNA-binding proteins or transcription factors. It is possible that certain ligands or disruptors could potentially affect TR binding to TREs vs. binding to other transcription factors. In support of this notion are the distinct phenotypes observed in knock-in mouse models that express mutant GR and TR that can no longer bind DNA but interact with other proteins vs. mice that do not express GR or TR (Reichardt et al. 1998; Shibusawa et al. 2003).

1.9 TR-Mediated Transcriptional Activity

TRs generally form heterodimers with another member of the nuclear receptor superfamily, the retinoid X receptor (RXR), and associate with corepressor and coactivator complexes on the promoters of target genes to modify histone acetylation and regulate their transcription, both in the absence and presence of TH, respectively (Fig. 1.2) (Aranda et al. 2013; Brent 2012; Oetting and Yen 2007). In the absence of TH, TR represses basal transcription of positively regulated target genes owing to TR binding to TREs and its recruitment of a corepressor complex that promotes histone deacetylation. In the presence of TH, basal repression is reversed and transcriptional activation occurs due to ligand-bound receptor recruitment of coactivator complexes. These processes are described in more detail below.

1.10 Basal Repression of Transcription by TRs

There is a family of corepressor proteins that include silencing mediator for retinoid and TH receptors (SMRT) and nuclear receptor corepressor (NCoR) that can bind to unliganded TRs and RARs and decrease transcription of positively regulated target genes (Brent 2012; Cheng et al. 2010; Oetting and Yen 2007). These corepressors are 270 kDa proteins that contain three transferable repression domains and two carboxy-terminal α-helical interaction domains with TR. They mediate basal repression by TR and retinoic acid receptor as well as by "orphan" members of the nuclear hormone receptor family such as rev-erbAα and chicken ovalbumin upstream transcription factor (COUP-TF) that do not bind to any identifiable ligands. Of note, corepressors do not interact with steroid hormone receptors and basal repression is not observed for these receptors. NCoR and SMRT contain two TR interaction sequences composed of consensus LXXI/HIXXXI/L sequences that resemble the LXXLL sequences of coactivators that enable their interaction with nuclear hormone receptors (Hu and Lazar 1999; Makowski et al. 2003). Interestingly, these respective motifs enable corepressors and coactivators to interact with similar amino acid residues on helices 3, 5, and 6 involved in the formation of the TR ligand-binding pocket. TH binding to TR results in conformational changes in these and other sites that distinguish corepressor vs. coactivator binding to TR.

Corepressors form a complex with other repressor proteins such as Sin 3 and histone deacetylases, particularly HDAC3 (Li et al. 2002). This complex causes histone deacetylation in the chromatin surrounding DNA regions near the TRE. These histone modifications lead to changes in the local chromatin structure that favor chromatin compaction and result in decreased recruitment of RNA pol II and general transcription factors and consequently repression of basal transcription. Methyl-CpG-binding proteins also can associate with a corepressor complex containing Sin3 and histone deacetylase to further increase basal repression in the absence of TH by increasing DNA methylation (Rietveld et al. 2002). T_3 binding to TR causes conformational changes in the TR that lead to dissociation of CoRs from TRs and recruitment of coactivators (CoAs) to the TRE-bound TR. Thus, T_3 reverses basal repression and stimulates transcription by recruiting different TR binding cofactors that increase histone acetylation near the TRE.

Studies in transgenic and knock-in models of dominant-negative NCoR mutant mice show that reduced NCoR action led to decreased basal repression and enhanced transcriptional activity in vivo (Astapova et al. 2008; Feng et al. 2001). The latter observation suggests that corepressors also can modulate T_3-mediated transcriptional activity by competing with coactivators for binding to TR despite having significantly lower binding affinity for ligand-bound TR than for unliganded TR. Histone deacetylase inhibitors such as trichostatin A and suberanilohydroxamic acid (SAHA/Vorinostat) can relieve basal repression and enhance transcription pharmacologically (Kim et al. 2014). Thus, it is possible that there may be environmental compounds that decrease basal repression and enhance TH-mediated transcription by interfering with corepressor binding to TR. This may be of particular significance in early pregnancy since the fetus is in a relatively privileged hypothyroid state due to increased placental *Dio 3* activity and the lack of a functioning fetal thyroid gland. Such disruptors of basal repression could have transcriptional effects on development and function of tissues where target genes need to be silenced during early pregnancy. Additionally, thyromimetic disruptors also could decrease corepressor binding to TRs during early development.

1.11 Transcriptional Activation by TRs

A significant number of transcriptional coactivators have been identified that interact with liganded nuclear hormone receptors and increase their transcriptional activity. However, it presently is not known whether these cofactors act only on specific genes or in specific tissues. Nevertheless, studies from many groups have shown that there are several major groups of coactivators such as the steroid receptor coactivators (SRCs), vitamin A receptor-interacting proteins/TR-associated proteins (DRIP/TRAPs), and CREB-binding protein (CBP)/p300 that play important roles in the transcription of many target genes regulated by nuclear hormone receptors (Brent 2012; Cheng et al. 2010; Oetting and Yen 2007).

SRCs are a family of 160 kDa proteins comprised of SRC-1, SRC-2, and SRC-3 that bind to ligand-bound nuclear hormone receptors such as TRs to promote ligand-dependent transcription. SRCs also interact with both CBP and the related E1A-interacting protein, p300 (Aranda et al. 2013; Cheng et al. 2010; Dasgupta et al. 2014; Oetting and Yen 2007). CBP/p300, in turn, interacts with PCAF (p300/CBP-associated factor), the mammalian homolog of the yeast transcriptional activator, general control non-repressed protein 5 (GCN5). Both CBP/p300 and PCAF have intrinsic histone acetyltransferase (HAT) activity.

DRIP/TRAPs also form a complex that interacts with ligand-bound VDR, TR, and other nuclear hormone receptors (Aranda et al. 2013; Brent 2012; Cheng et al. 2010; Dasgupta et al. 2014; Oetting and Yen 2007). It is noteworthy that none of the members of the DRIP/TRAP complex belong to the SRC family or any of the proteins that associate with them. The DRIP/TRAP subunits are mammalian homologs of the yeast Mediator complex, which associates with RNA Pol II to increase transcription. Current models suggest that ligand-bound TR recruits DRIP/TRAP complex and the latter recruits or stabilizes RNA Pol II complex by virtue of several common members. Of note, DRIP/TRAP complex does not have intrinsic HAT activity so it may primarily have an adapter function that links ligand-bound TRs with the general transcriptional machinery to activate transcription. Additionally, several studies have suggested that it may be recruited to TR after the SRC complex initiates changes in chromatin modification and structure. Indeed, chromatin immunoprecipitation assays of coactivator binding to TREs suggest that there is sequential, possibly cyclical, recruitment of coactivator complexes to TREs by TH-bound TRs. ATP-dependent chromatin remodeling proteins such as SW1/Snf and BRG-1 may be recruited early to TREs, followed by SRC complex and then DRIP/TRAP complex after TH binding to TRs.

Recently, TH was shown to metabolically activate Sirt-1 and induce deacetylation of transcription factors such as PGC1α and FOXO1 (Singh et al. 2013; Thakran et al. 2013). This deacetylation by Sirt1 was independent of its histone deacetylation activity. These findings support the notion that TH may be able to indirectly regulate a significant number of target genes without direct TR binding to their promoters through deacetylation of transcription factors by Sirt1. This mechanism would increase the repertoire of target genes regulated by TH without requiring TR binding to TREs or secondary transcription of other transcription factors. These findings also may provide an explanation for recent ChIP-seq findings that showed certain target genes activated by TH without any detectable TR binding in the promoter or other regions of the target gene (Ramadoss et al. 2014). Nuclear-cytoplasmic shuttling (Baumann et al. 2001), ubiquitination (Dace et al. 2000), acetylation (Lin et al. 2005), phosphorylation (Lin et al. 2003), and sumoylation (Liu et al. 2015) also play critical roles in TR stability and DNA binding.

1.12 Negative Transcriptional Regulation by TRs

A large number of target genes that are negatively regulated by THs have been identified by gene expression arrays of various tissues. These findings were initially surprising since few such target genes had been known previously. The precise mechanism(s) for negative regulation by TH are not well understood. In general, decreases in histone acetylation associated with negative regulation of target genes have been observed; however, exceptions also have been described (Ishii et al. 2004; Sasaki et al. 1999; Wang et al. 2009). It has been proposed that the latter may be due to changes at specific histone acetylation sites (rather than total histone acetylation) or from other histone modifications such as methylation or phosphorylation. The complexes that interact with TRs and the mechanism(s) involved in negative regulation currently need further investigation.

Several different mechanisms have been proposed to account for negative transcriptional regulation by TH (Santos et al. 2011). These explanations may apply in different situations or on different target genes. First, negative regulation may be due to TR interference with the actions of other transcription factors or basal transcription factors. For instance, TRs can inhibit the transcriptional activity of AP-1 or GATA1 on certain target genes (Matsushita et al. 2007). Second, negative regulation also can occur by direct binding of ligand-bound TR to specific TREs. In this connection, a negative TRE in the TSHβ gene has been reported. Lastly, ligand-bound TRs may potentially recruit cofactors utilized by other transcription factors, particularly other nuclear hormone receptors, which in turn, decrease the transcription of their target genes (squelching).

1.12.1 TR Action in the CNS

TRα1 is thought to play a prominent role in the CNS based on its high expression throughout the brain as it accounts for approximately 70–80 % of total T_3 binding (Ercan-Fang et al. 1996) and is detected early during embryonic neocortical development (Bradley et al. 1992). In contrast, expression of TRβ is much more restricted and mainly found postnatally in selected neuronal populations such as hippocampal pyramidal and granule cells, paraventricular hypothalamic neurons, and cerebellar Purkinje cells (Bradley et al. 1989; Strait et al. 1991). Investigation of receptor-deficient mice revealed that TRβ mediates TH effects during the development of retina photoreceptors and the auditory system (Jones et al. 2003).

Despite the widespread distribution of TRs in the CNS, mouse knockouts of specific TR isoforms or even all TH receptors do not mimic the strong phenotype of hypothyroid animals suggesting that the lack of the hormone (T_3) has more detrimental consequences than the loss of its receptors (Bernal 2007; Flamant and Samarut 2003; Forrest and Vennstrom 2000; O'Shea and Williams 2002; Yen et al. 2003). For instance, cerebellar development in TRα−1 null mice is not altered even when the animals are rendered hypothyroid (Morte et al. 2002). In contrast, mouse mutants expressing dominant-negative TH receptors due to mutations in the T_3-binding domain (TRβ Δ337T; TRalpha1R384C) exhibit deranged cerebellar Purkinje cell development (Hashimoto et al. 2001). These findings suggest that unliganded TRs may exhibit more severe effects by repressing CNS target genes than when TRs are absent and basal levels of transcription can still occur.

1.12.2 Molecular Targets of TRs in the CNS

Most attempts to identify target genes that are directly regulated by TH during brain development have had only limited success (Diez et al. 2008; Dong et al. 2005; Quignodon et al. 2007; Royland et al. 2008). In a recent study on TR binding sites and target genes using a ChIP-on-Chip approach in the developing mouse cerebellum (Dong et al. 2009), almost half of the identified TR binding sites did not have canonical TRE sequences. Furthermore, genes that contained a well-characterized TRE could be regulated by TH in a spatiotemporal manner. For instance, the expression of RC3/neurogranin, a neuron-specific calmodulin-binding protein which is important for synaptic plasticity and spatial learning, is controlled by TH only in the striatum but not in the cerebral cortex (Iniguez et al. 1996) indicating the presence of other unidentified regulatory mechanisms that spatially restrict TH-induced gene expression.

TH also exerts its action during brain development by controlling the expression of transcription factors that in turn regulate specific sets of genes that are important for neuronal differentiation. Examples of such transcription factors are BTEB (Denver and Williamson 2009), hairless (hr) (Potter et al. 2002), NeuroD (Chantoux and Francon 2002), RORalpha (Koibuchi and Chin 1998), and NGFI-A (Egr1) (Mellstrom et al. 1994). Besides directly regulating target genes at the transcriptional level, TH recently was shown to regulate genes in the brain posttranscriptionally via miRNAs (Dong et al. 2015). Additionally, TH regulation of DNA methylation of gene promoters (Sui and Li 2010) adds a further layer of complexity to its target gene regulation during brain development.

1.13 Non-genomic Actions of TH

THs as well as some of their metabolites have been reported to cause effects that do not depend upon transcriptional regulation by TR. These "non-genomic" effects can be mediated by proteins other than TRs that bind to TH or possibly by cytoplasmic TRs that cannot mediate transcriptional changes due to their intracellular partitioning (Fig. 1.3). Indeed, TRs have been shown to shuttle continuously between the cytoplasm and nucleus, although the fraction residing in the cytoplasm is less than 10 % (Maruvada et al. 2003). Several lines of evidence that have been used to argue for non-genomic and non-TR effects of TH are rapid time course of action (faster than transcription and protein synthesis), utilization of membrane-signaling pathways such as kinases or calmodulin, occurrence that is independent of TR expression, and atypical affinity-activity relationships by TH and its analogs that are different than those observed for nuclear TRs (Cheng et al. 2010). Some of the nonnuclear sites for TH binding and action that have been described so far are plasma membrane-associated T_3 transporters, calcium ATPase, adenylate cyclase, and glucose transporters; an endoplasmic reticulum associated protein, prolyl hydroxylase; and monomeric pyruvate kinase (Cheng et al. 2010).

TH also has major effects on mitochondrial activity and cellular energy state. In this connection, a 43 kDa mitochondrial protein has been described that bound to TREs and could be recognized by antibodies against the TRα LBD, suggesting that it might be a TR variant (Casas et al. 1999; Sinha et al. 2010). TRβs interact with the p85 subunit of PI3K and activate the PI3K-Akt/PKB signaling cascade; thus, the small subpopulation of cytosolic TRβ may be involved in cell signaling (Cheng et al. 2010). Rapid effects of TH on PI3K signaling in the neuron also have been described.

Davis and others have shown that T_4 and a TH analog, TRIAC, bind to integrin α-Vβ-3 with higher affinity than T_3 (Cohen et al. 2011). Their binding to this plasma membrane-bound integrin led to activation of the MAPK cascade. This alternative pathway for TH action raises the possibility that endocrine disruptors may be able to activate cell surface proteins such as integrin and modulate TH action without necessarily entering into the cell. In yet another mode of non-genomic action, T_4 and its metabolite, reverse T_3 (but not T_3), profoundly stimulate actin polymerization in cultured astrocytes and cerebellar granule cells thereby promoting neuronal outgrowth (Farwell et al. 2006).

1.14 Summary

TH action is regulated at many different levels – secretion, transport, cellular uptake, deiodinase activity, and TR and cofactor expression as well as transcriptional, post-transcriptional, and non-genomical. Each of these steps of TH action can potentially be affected by endocrine disruptors with serious consequences on the development and function of the CNS. Additionally, there appears to be critical spatiotemporal aspects of TH action as unliganded TRs repress transcription of some genes in the CNS early in pregnancy, and liganded receptors regulate target gene expression later in pregnancy. These effects are not only controlled by TR expression and ligand availability but also by tissue-specific and temporal epigenetic effects such as DNA methylation, histone modifications, and miRNAs (Fig. 1.3). Thus, where and when a fetus or adult organism is exposed to endocrine disruptors may play critical roles in determining whether the disruptors may have deleterious effects. In this textbook, we will examine in more detail some examples of endocrine disruption of TH action in the CNS.

References

Aranda A, Alonso-Merino E, Zambrano A (2013) Receptors of thyroid hormones. Pediatr Endocrinol Rev 11:2–13

Astapova I, Lee LJ, Morales C, Tauber S, Bilban M, Hollenberg AN (2008) The nuclear corepressor, NCoR, regulates thyroid hormone action in vivo. Proc Natl Acad Sci U S A 105:19544–19549

Baumann CT, Maruvada P, Hager GL, Yen PM (2001) Nuclear cytoplasmic shuttling by thyroid hormone receptors. multiple protein interactions are required for nuclear retention. J Biol Chem 276:11237–11245

Bernal J (2007) Thyroid hormone receptors in brain development and function. Nat Clin Pract Endocrinol Metab 3:249–259

Bockmann J, Winter C, Wittkowski W, Kreutz MR, Bockers TM (1997) Cloning and expression of a brain-derived TSH receptor. Biochem Biophys Res Commun 238:173–178

Bradley DJ, Young WS III, Weinberger C (1989) Differential expression of alpha and beta thyroid hormone receptor genes in rat brain and pituitary. Proc Natl Acad Sci U S A 86:7250–7254

Bradley DJ, Towle HC, Young WS III (1992) Spatial and temporal expression of alpha- and beta-thyroid hormone receptor mRNAs, including the beta 2-subtype, in the developing mammalian nervous system. J Neurosci 12:2288–2302

Brent GA (2012) Mechanisms of thyroid hormone action. J Clin Invest 122:3035–3043

Burger AG, Lambert M, Cullen M (1981) Interference in the conversion of T4 to T3 and rT3 by medications in man. Ann Endocrinol 42:461–469

Casas F, Rochard P, Rodier A, Cassar-Malek I, Marchal-Victorion S, Wiesner RJ, Cabello G, Wrutniak C (1999) A variant form of the nuclear triiodothyronine receptor c-ErbAalpha1 plays a direct role in regulation of mitochondrial RNA synthesis. Mol Cell Biol 19:7913–7924

Chantoux F, Francon J (2002) Thyroid hormone regulates the expression of NeuroD/BHF1 during the development of rat cerebellum. Mol Cell Endocrinol 194:157–163

Cheek AO, Kow K, Chen J, McLachlan JA (1999) Potential mechanisms of thyroid disruption in humans: interaction of organochlorine compounds with thyroid receptor, transthyretin, and thyroid-binding globulin. Environ Health Perspect 107:273–278

Cheng SY, Leonard JL, Davis PJ (2010) Molecular aspects of thyroid hormone actions. Endocr Rev 31:139–170

Cohen K, Ellis M, Khoury S, Davis PJ, Hercbergs A, Ashur-Fabian O (2011) Thyroid hormone is a MAPK-dependent growth factor for human myeloma cells acting via alphavbeta3 integrin. Mol Cancer Res 9:1385–1394

Dace A, Zhao L, Park KS, Furuno T, Takamura N, Nakanishi M, West BL, Hanover JA, Cheng S (2000) Hormone binding induces rapid proteasome-mediated degradation of thyroid hormone receptors. Proc Natl Acad Sci U S A 97:8985–8990

Dasgupta S, Lonard DM, O'Malley BW (2014) Nuclear receptor coactivators: master regulators of human health and disease. Annu Rev Med 65:279–292

Denver RJ, Williamson KE (2009) Identification of a thyroid hormone response element in the mouse Kruppel-like factor 9 gene to explain its postnatal expression in the brain. Endocrinology 150:3935–3943

Diez D, Grijota-Martinez C, Agretti P, De Marco G, Tonacchera M, Pinchera A, de Escobar GM, Bernal J, Morte B (2008) Thyroid hormone action in the adult brain: gene expression profiling of the effects of single and multiple doses of triiodo-L-thyronine in the rat striatum. Endocrinology 149:3989–4000

Dong H, Wade M, Williams A, Lee A, Douglas GR, Yauk C (2005) Molecular insight into the effects of hypothyroidism on the developing cerebellum. Biochem Biophys Res Commun 330:1182–1193

Dong H, Yauk CL, Rowan-Carroll A, You SH, Zoeller RT, Lambert I, Wade MG (2009) Identification of thyroid hormone receptor binding sites and target genes using ChIP-on-chip in developing mouse cerebellum. PLoS One 4:e4610

Dong H, You SH, Williams A, Wade MG, Yauk CL, Thomas Zoeller R (2015) Transient maternal hypothyroxinemia potentiates the transcriptional response to exogenous thyroid hormone in the fetal cerebral cortex before the onset of fetal thyroid function: a messenger and microRNA profiling study. Cereb Cortex 25(7):1735–1745

Dumitrescu AM, Refetoff S (2013) The syndromes of reduced sensitivity to thyroid hormone. Biochim Biophys Acta 1830:3987–4003

Ercan-Fang S, Schwartz HL, Oppenheimer JH (1996) Isoform-specific 3,5,3′-triiodothyronine receptor binding capacity and messenger ribonucleic acid content in rat adenohypophysis: effect of thyroidal state and comparison with extrapituitary tissues. Endocrinology 137:3228–3233

Farwell AP, Dubord-Tomasetti SA, Pietrzykowski AZ, Leonard JL (2006) Dynamic nongenomic actions of thyroid hormone in the developing rat brain. Endocrinology 147:2567–2574

Fekete C, Lechan RM (2014) Central regulation of hypothalamic-pituitary-thyroid axis under physiological and pathophysiological conditions. Endocr Rev 35:159–194

Feng X, Jiang Y, Meltzer P, Yen PM (2001) Transgenic targeting of a dominant negative corepressor to liver blocks basal repression by thyroid hormone receptor and increases cell proliferation. J Biol Chem 276:15066–15072

Flamant F, Samarut J (2003) Thyroid hormone receptors: lessons from knockout and knock-in mutant mice. Trends Endocrinol Metab 14:85–90

Forrest D, Vennstrom B (2000) Functions of thyroid hormone receptors in mice. Thyroid 10:41–52

Garcia M, Fernandez A, Moreno JC (2014) Central hypothyroidism in children. Endocr Dev 26:79–107

Grasberger H, Refetoff S (2011) Genetic causes of congenital hypothyroidism due to dyshormonogenesis. Curr Opin Pediatr 23:421–428

Hashimoto K, Curty FH, Borges PP, Lee CE, Abel ED, Elmquist JK, Cohen RN, Wondisford FE (2001) An unliganded thyroid hormone receptor causes severe neurological dysfunction. Proc Natl Acad Sci U S A 98:3998–4003

Heuer H, Visser TJ (2013) The pathophysiological consequences of thyroid hormone transporter deficiencies: Insights from mouse models. Biochim Biophys Acta 1830:3974–3978

Hu X, Lazar MA (1999) The CoRNR motif controls the recruitment of corepressors by nuclear hormone receptors. Nature 402:93–96

Iniguez MA, De Lecea L, Guadano-Ferraz A, Morte B, Gerendasy D, Sutcliffe JG, Bernal J (1996) Cell-specific effects of thyroid hormone on RC3/neurogranin expression in rat brain. Endocrinology 137:1032–1041

Ishii S, Yamada M, Satoh T, Monden T, Hashimoto K, Shibusawa N, Onigata K, Morikawa A, Mori M (2004) Aberrant dynamics of histone deacetylation at the thyrotropin-releasing hormone gene in resistance to thyroid hormone. Mol Endocrinol 18:1708–1720

Jones I, Srinivas M, Ng L, Forrest D (2003) The thyroid hormone receptor beta gene: structure and functions in the brain and sensory systems. Thyroid 13:1057–1068

Kim DW, Park JW, Willingham MC, Cheng SY (2014) A histone deacetylase inhibitor improves hypothyroidism caused by a TRalpha1 mutant. Hum Mol Genet 23:2651–2664

Koibuchi N, Chin WW (1998) ROR alpha gene expression in the perinatal rat cerebellum: ontogeny and thyroid hormone regulation. Endocrinology 139:2335–2341

Larsen PR, Zavacki AM (2012) The role of the iodothyronine deiodinases in the physiology and pathophysiology of thyroid hormone action. Eur Thyroid J 1:232–242

Li J, Lin Q, Wang W, Wade P, Wong J (2002) Specific targeting and constitutive association of histone deacetylase complexes during transcriptional repression. Genes Dev 16:687–692

Lin HY, Zhang S, West BL, Tang HY, Passaretti T, Davis FB, Davis PJ (2003) Identification of the putative MAP kinase docking site in the thyroid hormone receptor-beta1 DNA-binding domain: functional consequences of mutations at the docking site. Biochemistry 42:7571–7579

Lin HY, Hopkins R, Cao HJ, Tang HY, Alexander C, Davis FB, Davis PJ (2005) Acetylation of nuclear hormone receptor superfamily members: thyroid hormone causes acetylation of its own receptor by a mitogen-activated protein kinase-dependent mechanism. Steroids 70:444–449

Liu YY, Ayers S, Milanesi A, Teng X, Rabi S, Akiba Y, Brent GA (2015) Thyroid hormone receptor sumoylation is required for preadipocyte differentiation and proliferation. J Biol Chem 290:7402–7415

Makowski A, Brzostek S, Cohen RN, Hollenberg AN (2003) Determination of nuclear receptor corepressor interactions with the thyroid hormone receptor. Mol Endocrinol 17:273–286

Maruvada P, Baumann CT, Hager GL, Yen PM (2003) Dynamic shuttling and intranuclear mobility of nuclear hormone receptors. J Biol Chem 278:12425–12432

Matsushita A, Sasaki S, Kashiwabara Y, Nagayama K, Ohba K, Iwaki H, Misawa H, Ishizuka K, Nakamura H (2007) Essential role of GATA2 in the negative regulation of thyrotropin beta gene by thyroid hormone and its receptors. Mol Endocrinol 21:865–884

Mellstrom B, Pipaon C, Naranjo JR, Perez-Castillo A, Santos A (1994) Differential effect of thyroid hormone on NGFI-A gene expression in developing rat brain. Endocrinology 135:583–588

Morte B, Manzano J, Scanlan T, Vennstrom B, Bernal J (2002) Deletion of the thyroid hormone receptor alpha 1 prevents the structural alterations of the cerebellum induced by hypothyroidism. Proc Natl Acad Sci U S A 99:3985–3989

O'Shea PJ, Williams GR (2002) Insight into the physiological actions of thyroid hormone receptors from genetically modified mice. J Endocrinol 175:553–570

Oetting A, Yen PM (2007) New insights into thyroid hormone action. Best Pract Res Clin Endocrinol Metab 21:193–208

Potter GB, Zarach JM, Sisk JM, Thompson CC (2002) The thyroid hormone-regulated corepressor hairless associates with histone deacetylases in neonatal rat brain. Mol Endocrinol 16:2547–2560

Quignodon L, Grijota-Martinez C, Compe E, Guyot R, Allioli N, Laperriere D, Walker R, Meltzer P, Mader S, Samarut J et al (2007) A combined approach identifies a limited number of new thyroid hormone target genes in post-natal mouse cerebellum. J Mol Endocrinol 39:17–28

Ramadoss P, Abraham BJ, Tsai L, Zhou Y, Costa-e-Sousa RH, Ye F, Bilban M, Zhao K, Hollenberg AN (2014) Novel mechanism of positive versus negative regulation by thyroid hormone receptor beta1 (TRbeta1) identified by genome-wide profiling of binding sites in mouse liver. J Biol Chem 289:1313–1328

Reader SC, Carroll B, Robertson WR, Lambert A (1987) Assessment of the biopotency of antithyroid drugs using porcine thyroid cells. Biochem Pharmacol 36:1825–1828

Reichardt HM, Kaestner KH, Tuckermann J, Kretz O, Wessely O, Bock R, Gass P, Schmid W, Herrlich P, Angel P et al (1998) DNA binding of the glucocorticoid receptor is not essential for survival. Cell 93:531–541

Rietveld LE, Caldenhoven E, Stunnenberg HG (2002) In vivo repression of an erythroid-specific gene by distinct corepressor complexes. EMBO J 21:1389–1397

Royland JE, Parker JS, Gilbert ME (2008) A genomic analysis of subclinical hypothyroidism in hippocampus and neocortex of the developing rat brain. J Neuroendocrinol 20:1319–1338

Santos GM, Fairall L, Schwabe JW (2011) Negative regulation by nuclear receptors: a plethora of mechanisms. Trends Endocrinol Metab 22:87–93

Sasaki S, Lesoon-Wood LA, Dey A, Kuwata T, Weintraub BD, Humphrey G, Yang WM, Seto E, Yen PM, Howard BH et al (1999) Ligand-induced recruitment of a histone deacetylase in the negative-feedback regulation of the thyrotropin beta gene. EMBO J 18:5389–5398

Schussler GC (2000) The thyroxine-binding proteins. Thyroid 10:141–149

Shibusawa N, Hashimoto K, Nikrodhanond AA, Liberman MC, Applebury ML, Liao XH, Robbins JT, Refetoff S, Cohen RN, Wondisford FE (2003) Thyroid hormone action in the absence of thyroid hormone receptor DNA-binding in vivo. J Clin Invest 112:588–597

Singh BK, Sinha RA, Zhou J, Xie SY, You SH, Gauthier K, Yen PM (2013) FoxO1 deacetylation regulates thyroid hormone-induced transcription of key hepatic gluconeogenic genes. J Biol Chem 288:30365–30372

Sinha RA, Pathak A, Mohan V, Babu S, Pal A, Khare D, Godbole MM (2010) Evidence of a bigenomic regulation of mitochondrial gene expression by thyroid hormone during rat brain development. Biochem Biophys Res Commun 397:548–552

Smith TJ (2015) TSH-receptor-expressing fibrocytes and thyroid-associated ophthalmopathy. Nat Rev Endocrinol 11:171–181

Strait KA, Schwartz HL, Seybold VS, Ling NC, Oppenheimer JH (1991) Immunofluorescence localization of thyroid hormone receptor protein beta 1 and variant alpha 2 in selected tissues: cerebellar Purkinje cells as a model for beta 1 receptor-mediated developmental effects of thyroid hormone in brain. Proc Natl Acad Sci U S A 88:3887–3891

Sui L, Li BM (2010) Effects of perinatal hypothyroidism on regulation of reelin and brain-derived neurotrophic factor gene expression in rat hippocampus: role of DNA methylation and histone acetylation. Steroids 75:988–997

Thakran S, Sharma P, Attia RR, Hori RT, Deng X, Elam MB, Park EA (2013) Role of sirtuin 1 in the regulation of hepatic gene expression by thyroid hormone. J Biol Chem 288:807–818

Wang D, Xia X, Liu Y, Oetting A, Walker RL, Zhu Y, Meltzer P, Cole PA, Shi YB, Yen PM (2009) Negative regulation of TSHalpha target gene by thyroid hormone involves histone acetylation and corepressor complex dissociation. Mol Endocrinol 23:600–609

Yen PM, Feng X, Flamant F, Chen Y, Walker RL, Weiss RE, Chassande O, Samarut J, Refetoff S, Meltzer PS (2003) Effects of ligand and thyroid hormone receptor isoforms on hepatic gene expression profiles of thyroid hormone receptor knockout mice. EMBO Rep 4:581–587

Chapter 2
Deiodinase and Brain Development

Masami Murakami

Abstract Thyroid hormone receptors are enriched in neurons which are the primary target of T_3 actions. Type 2 iodothyronine deiodinase (D2), which catalyzes conversion of thyroxine (T_4) to 3,5,3'-triiodothyronine (T_3), is expressed predominantly in glial cells, and type 3 iodothyronine deiodinase (D3), which catalyzes conversion of T4 to 3,3',5'-triiodothyronine (rT_3) and T3 to 3,3'-diiodothyronine (T_2), is expressed in neurons. Thyroid hormone metabolism by D2 and D3 plays important roles in brain function and development.

Keywords Thyroid hormone • Iodothyronine deiodinase • Thyroid hormone receptor • Brain • Neuron • Astrocyte • Development

2.1 Introduction

Thyroid hormones play important roles in vertebrate physiology and development, including fetal and postnatal nervous system development and the maintenance of adult brain function (Morreale de Escobar et al. 2004). Severe thyroid hormone deficiency in fetal and neonatal periods results in cretinism, a disease characterized by mental retardation, deafness, and ataxia, which are irreversible if not treated with thyroid hormone soon after birth. Congenital hypothyroidism is usually diagnosed within the first weeks of life by a neonatal TSH screening test (Horn and Heuer 2010; Schroeder and Privalsky 2014). Untreated hypothyroidism in the adult is associated with severe intellectual defects, abnormal balance and defects in fine motor skills, spasticity, and deafness (DeLong et al. 1985). Correcting thyroid hormone deficiencies is critical for normal brain development and function.

The thyroid gland synthesizes and secretes thyroid hormones, thyroxine (T_4) and 3,5,3'-triiodothyronine (T_3). In order to bind to thyroid hormone receptors (TRs) and exert its biological activity, T_4, which is the major secretory product of the

M. Murakami (✉)
Department of Clinical Laboratory Medicine, Gunma University Graduate School of Medicine, 3-39-15 Showa-machi, Maebashi 371-8511, Japan
e-mail: mmurakam@gunma-u.ac.jp

© Springer Science+Business Media New York 2016

N. Koibuchi, P.M. Yen (eds.), *Thyroid Hormone Disruption and Neurodevelopment*, Contemporary Clinical Neuroscience, DOI 10.1007/978-1-4939-3737-0_2

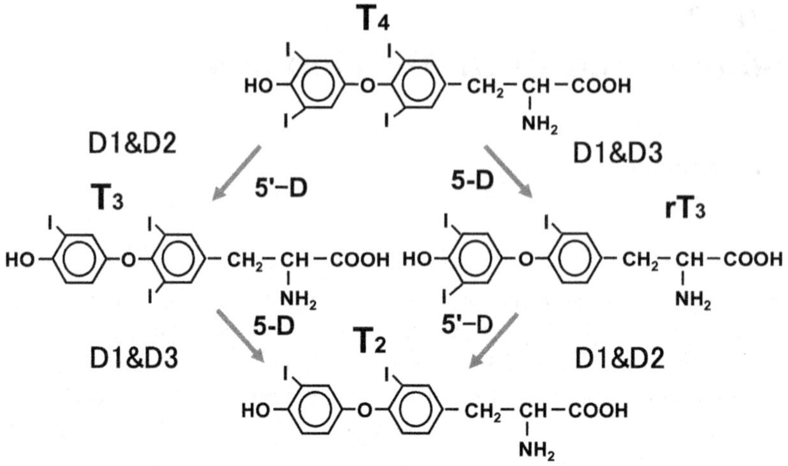

Fig. 2.1 Metabolism of thyroid hormones by iodothyronine deiodinases (D1, D2, D3)

thyroid gland, needs to be converted to T_3 by selenocysteine containing oxidoreductases, namely, iodothyronine deiodinases (Bianco et al. 2002).

Regulation of thyroid hormone metabolism by iodothyronine deiodinases plays a pivotal role in brain development and function (Darras et al. 2015).

2.1.1 Thyroid Hormone Metabolism

There are three types of iodothyronine deiodinase: type 1 (D1), type 2 (D2), and type 3 (D3). D1 and D2 remove iodine from the outer ring of T_4 to form the active thyroid hormone T_3. D1 and D3 remove iodine from the inner rings of T_4 and T_3 to form the inactive thyroid hormones $3,3',5'$-triiodothyronine (rT_3) and $3,3'$-diiodothyronine (T_2), respectively (Fig. 2.1). The characteristics of iodothyronine deiodinases are summarized in Table 2.1. D1 is present in the thyroid gland, liver, kidney, and many other tissues, whereas D2 is present in a limited number of tissues, including the central nervous system, anterior pituitary, and brown fat in the rat. In humans, D2 has also been reported to exist in the thyroid (Murakami et al. 2001), skeletal muscle (Hosoi et al. 1999), and vascular smooth muscle (Mizuma et al. 2001). D3 is present in the placenta, uterus, skin, central nervous system, and fetal liver. Although both D1 and D2 catalyze conversion of T_4 to T_3, the properties of these two enzymes are remarkably different. The *Km* value of D2 is approximately 1–2 nM for T_4, which is 100-fold lower than that of D1. D1, but not D2, is highly sensitive to inhibition by the antithyroid drug, 6-n-propylthiouracil (PTU). D1 activity is known to decrease in a hypothyroid state and is believed to have a primary role in maintaining circulating T_3 levels. D2 activity, in contrast, is elevated in a hypothyroid state and is considered to play a pivotal role in providing local T_3 to

Table 2.1 Characteristics of iodothyronine deiodinases

Parameter	D1 (5′-D, 5-D)	D2 (5′-D)	D3 (5-D)
Biochemical properties			
Molecular mass (kDa)	29	31	32
Substrate preference	$rT_3 \gg T_4 > T_3$	$T_4 > rT_3$	$T_3 > T_4$
Apparent Km	0.1–10 μM	1–2 nM	5–20 nM
Protein half-life	Hours	~20 min	Hours
Subcellular localization	Plasma membrane	Endoplasmic reticulum	Plasma membrane
Homodimerization	Yes	Yes	Yes
Chromosomal	1p32-p33	14q24.3	14q32
Susceptibility to PTU	High	Very low	Very low
Susceptibility to IOP	Yes	Yes	Yes
Tissue distribution	Liver, kidney,	Brain, pituitary	Brain, skin, placenta
	Thyroid	Brown adipose tissue, thyroid[a]	Uterus, fetus
		Skeletal muscle[a], vascular smooth muscle[a]	
Response to thyroid hormone			
Transcriptional	↑↑	↓	↑↑
Posttranslational	?	↓↓↓ (ubiquitination)	?
Physiological role	Major source of plasma T_3	Provides intracellular T_3	T_3, T_4 clearance

[a]Observed in human tissue

regulate intracellular T_3 concentration. D3 activity is insensitive to inhibition by PTU and decreases in a hypothyroid state (Bianco et al. 2002; Gereben et al. 2008; St. Germain et al. 2009) (Table 2.1).

2.1.2 Iodothyronine Deiodinases and Central Nervous System

Studies in animal models have revealed that approximately 80 % of T_3 is produced locally in the central nervous system suggesting an important role of D2 which catalyzes outer ring deiodination (Crantz et al. 1982).

TRs are enriched in neurons which are the primary target of T_3 actions. In contrast, it has been demonstrated that D2 is expressed predominantly in glial cells: the astrocytes throughout the brain (Guadano-Ferraz et al. 1997; Murakami et al. 2000) and in the tanycytes lining part of the third ventricle surface (Tu et al. 1997). The *Dio2* mRNA is not restricted to the cell body but is also present along the cellular processes. These results indicate a crucial role for glial cells in thyroid hormone metabolism in the brain and a close coupling between glial cells and neurons in thyroid hormone homeostasis. Circulating T_4 and T_3 enter the brain through the

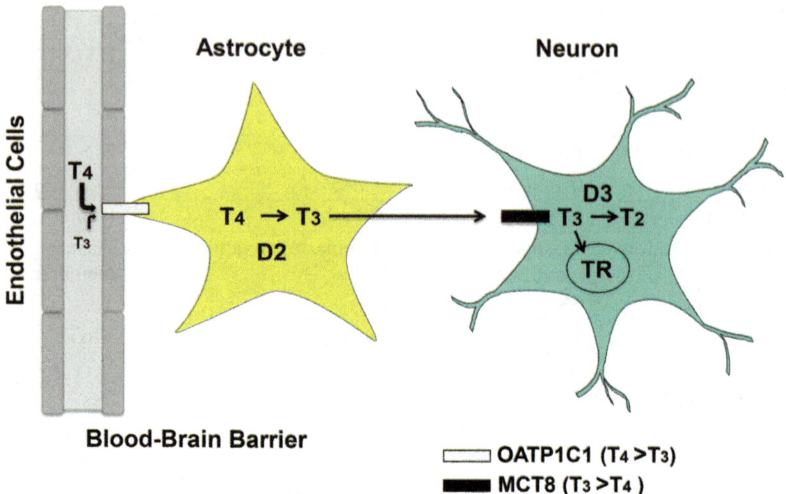

Fig. 2.2 Thyroid hormone metabolism in astrocytes and neurons

blood-brain barrier by the OATP1C1 transporter. T_4 reaches the astrocytes through their end-feet in contact with the capillaries and produce additional T_3 by outer ring deiodination by D2. Paracrine interaction between astrocytes and neurons is demonstrated by an in vitro coculture system of glioma cells and neuroblastoma cells (Freitas et al. 2010). In the presence of T_4, D2 activity in glial cells resulted in increased T_3 production that reached neurons through the MCT8 transporter and promoted thyroid hormone-responsive gene expression (Horn and Heuer 2010; Morte and Bernal 2014) (Fig. 2.2).

Activity of D2 is highly regulated by the thyroid status via both pretranslational mechanisms and posttranslational mechanisms through ubiquitination. Hypothyroidism results in upregulated D2 activities and hyperthyroidism leads to a decrease in D2 activities (Table 2.1). These regulation mechanisms by the thyroid status are interpreted as a protective mechanism to maintain the brain T_3 content in light of altered circulating thyroid hormone levels. D3 is another important factor for thyroid hormone metabolism in the brain. D3 is strongly expressed in neurons where it inactivates both T_4 and T_3 by inner ring deiodination to rT_3 and T_2 so as to downregulate local thyroid hormone concentrations and protect the neuron from supraphysiological levels of thyroid hormone (Bianco et al. 2002; Horn and Heuer 2010).

Astrocytes generate active T_3 from circulating pro-hormone T_4, whereas neurons degrade both T_4 and T_3 to inactive rT_3 and T_2, respectively, and thereby regulate local thyroid hormone availability within the brain. This balancing act protects the brain from the detrimental effects of hyper- or hypothyroidism (Morte and Bernal 2014; Schroeder and Privalsky 2014).

2.1.3 Thyroid Hormones in Neurological Development

The fetal thyroid gland reaches maturity by week 11–12, close to the end of the first trimester, and begins to secrete thyroid hormones by about week 16 (Obregon et al. 2007). During this period, an adequate supply of maternal thyroid hormones is required to ensure normal neurological development. Hypothyroid fetuses suffer postnatal disorders including mental retardation, deafness, and spasticity. Severe iodine deficiency, which causes both maternal and fetal hypothyroidism, is the most common cause of mental retardation (Glinoer 2001; Morreale de Escobar et al. 2004; Pearce 2009). Recent evidence suggests that even mild reduction in maternal thyroid hormone levels in early pregnancy is associated with reduced IQ in offspring (LaFranchi and Austin 2007; Patel et al. 2011).

Thyroid hormones act on embryological and fetal tissues early in development. Thyroid hormone and associated receptors are present in human fetal tissues prior to the production and secretion of thyroid hormones from fetal thyroid gland at 16–18 weeks of gestation, as demonstrated by detection of T_4 and T_3 in the human cerebral cortex by week 12 gestation (Calvo et al. 2002; Kester et al. 2004). Active transport of maternal thyroid hormone across the placenta is occurring during this critical period of gestation, and transplacental thyroid hormone transfer from maternal to fetal circulation to ensure appropriate fetal thyroid hormone levels is important in CNS development (Bernal 2007). Following onset of active T_4 secretion by the fetus, T_4 levels in fetal tissues parallel those in fetal plasma. In contrast to tissue T_4 levels, T_3 concentration varies in different tissues. T_3 levels are low in fetal liver and plasma and high in brain and brown adipose tissue (Obregon et al. 2007). These differences are attributed to variations in the activity of D2, which suggests an important role for T_3 in brain development and maturation. D3 is active in the placenta and fetal tissues, ensuring the fetus is not exposed to excessive amounts of maternal T_4 (Galton 2005).

Normal neurological development also depends on thyroid hormone in rodents. TRs and iodothyronine deiodinases are expressed in the early brain before the thyroid gland develops (Obregon et al. 2007). The critical period for thyroid hormone action in rat brain is estimated to extend from around embryonic day 18 (E18) to postnatal day 21–25 (P21–25) (Porterfield and Hendrich 1993). Abnormalities in brain development in hypothyroid rats are seen in the postnatal period and are demonstrated by reduced maturation of structures, such as the cerebellum, where delayed granular cell migration and Purkinje cell maturation are prevented (Koibuchi et al. 2003). Although this may suggest that rat brain is more affected by thyroid hormone action after birth, it should be recognized that TRs are seen very early in rat development. The developing brain is dependent on a supply of T_4, which is locally converted to T_3 by D2, and replacement with T_3 does not adequately replenish brain T_3 levels (Calvo et al. 1990). This emphasizes the need for maternal T_4 levels to be maintained to ensure normal fetal brain development. This also explains why even minimally reduced maternal T_4 levels in early pregnancy results in adverse outcomes to the offspring (Lavado-Autric et al. 2003; Auso et al. 2004).

2.1.4 Iodothyronine Deiodinases in Neurological Development

In the rat, D2 expression is first detectable at E16.5 and increases successively until postnatal day 15. Ontogeny of D2 in the fetal human brain demonstrated the occurrence of D2 in the developing cerebral cortex during the first trimester of pregnancy when the cortical T_3 concentration can first be detected (Chan et al. 2002). The similar ontogenic profile of D2 expression and T_3 content in different developing brain structures suggests that D2 is crucial in providing developing brain structures with T_3 produced from maternally derived T_4 (Horn and Heuer 2010). As mentioned above, D3 is expressed in fetal tissues and the placenta where it initially prevents maternal thyroid hormone access to the developing fetus (Williams 2008).

D2 knockout (KO) mice have normal plasma T_3 but increased levels of T_4 and TSH as a result of defect of T_4 to T_3 conversion in the anterior pituitary (Schneider et al. 2001). D3KO mice demonstrate a complicated pattern of thyroid function. They show a severalfold increase in plasma T_3 both in utero and in the neonatal period, due to a reduced T_3 clearance rate (Hernandez et al. 2006). However, from postnatal day 15 up to adulthood, they demonstrate central hypothyroidism with low plasma T_3 and T_4 levels, suggesting that the neonatal thyrotoxicosis has disturbed the maturation of the hypothalamus-pituitary-thyroid axis (Hernandez et al. 2006).

Although D2KO and D3KO mice did not show a severe neurological phenotype initially, detailed studies revealed the impaired ability to invert and descend a vertical pole (Galton et al. 2007; Peeters et al. 2013). The possible link of iodothyronine deiodinase deficiency with the structural development of the ear, eye, and brain has been studied to a large extent during the first postnatal weeks, consistent with the relatively late maturation of the rodent brain and sensory systems. Expression of *Dio3* is high in the immature prenatal cochlea and decreases in the first days after birth. In contrast, *Dio2* expression is very low in the prenatal cochlea but increases strikingly in the first postnatal week, prior to the onset of hearing in the second week (Ng et al. 2013). D2 deficiency delays cochlear development, while D3 deficiency accelerates the process. However, both conditions lead to impaired development of the sensory epithelium and permanent deafness (Ng et al. 2004, 2009). The cellular expression pattern of D3 partly overlaps with that of TRβ, the receptor that is primarily responsible for auditory development (Ng et al. 2009). These indicate the importance of local changes in the balance between thyroid hormone activation and inactivation and the fact that these changes have to occur in a critical time window to ensure normal development. Appropriate T_3 signaling is also crucial for eye development. D3 activity levels are high in prenatal mouse retina while no D2 activity is detected, suggesting that D3 has a protective role at these stages (Ng et al. 2010). The number of cones is greatly reduced in the retina of adult D3KO mice. This has been linked to an excessive T_3 signaling via TRβ2, leading to a transient increase in cell death of cones in the first postnatal days (Ng et al. 2010).

In the first days after birth, T_3 content is significantly increased in D3KO mouse brain, resulting in increased expression of thyroid hormone-responsive genes such as *Hairless* and *RC3* (Hernandez et al. 2006). By two weeks after birth, when peripheral thyroid hormone levels decreased, some brain regions still have increased T_3 levels, while T_3 levels in some other regions have already decreased, indicating that D3 has a region-specific role in regulating T_3 signaling in the developing brain (Hernandez et al. 2010). T_3 content in the neonatal D2KO mouse brain is markedly reduced. However, expression of several thyroid hormone-responsive genes is not or only modestly affected, suggesting that compensatory mechanisms exist, limiting the influence of D2 deficiency (Galton et al. 2007). The impact of D2 and D3 deficiency is higher on negatively regulated thyroid hormone-responsive genes than on positively regulated genes in the study on the expression of cerebral cortex genes in 4-week-old mice (Hernandez et al. 2012). A recent study on cerebellum shows that D2 activity is very low in the late fetal stages and at birth but increases during the first postnatal week, while D3 activity is relatively high prenatally and starts to decrease 3 days after birth (Peeters et al. 2013). D3KO mice show several abnormalities in cerebellar development, including reduced foliation, accelerated disappearance of the external germinal layer, and premature expansion of the molecular layer. Combined deletion of the TRα1 substantially corrects the cerebellar phenotype but, for instance, not the deafness of D3KO mice where TRβ signaling is important. This illustrates that the combined expression of D3 and specific receptor isoforms is controlling development of different tissues (Peeters et al. 2013).

References

Auso E, Lavado-Autric R, Cuevas E, Escobar del Rey F et al (2004) A moderate and transient deficiency of maternal thyroid function at the beginning of fetal neocorticogenesis alters neuronal migration. Endocrinology 145:4037–4047

Bernal J (2007) Thyroid hormone receptors in brain development and function. Nat Clin Pract Endocrinol Metab 3:249–259

Bianco AC, Salvatore D, Gereben B et al (2002) Biochemistry, cellular and molecular biology, and physiological roles of the iodothyronine selenodeiodinases. Endocr Rev 23:38–89

Calvo R, Obregon MJ, Ruiz de Ona C et al (1990) Congenital hypothyroidism, as studied in rats. Crucial role of maternal thyroxine but not 3,5,3′-triiodothyronine in the protection of the fetal brain. J Clin Invest 86:889–899

Calvo RM, Jauniaux E, Gulbis B et al (2002) Fetal tissues are exposed to biologically relevant free thyroxine concentrations during early phases of development. J Clin Endocrinol Metab 87:1768–1777

Chan S, Kachilele S, McCabe CJ et al (2002) Early expression of thyroid hormone deiodinase and receptors in human fetal brain cortex. Brain Res Dev Brain Res 138:109–116

Crantz FR, Silva JE, Larsen PR (1982) An analysis of the sources and quality of 3,5,3′-triiodothyronine specifically bound to nuclear receptors in rat cerebral cortex and cerebellum. Endocrinology 110:367–375

Darras VM, Houbrechts AM, Van Herck SLJ (2015) Intracellular thyroid hormone metabolism as a local regulator of nuclear thyroid hormone receptor-mediated impact on vertebrate development. Biochim Biophys Acta 1849(2):130–141. doi:10.1016/j.bbagrm.2014.05.004

DeLong GR, Stanbury JB, Fierro-Benitez R (1985) Neurological signs in congenital iodine-deficiency disorder (endemic cretinism). Dev Med Child Neurol 27:317–324

Freitas BC, Gereben B, Castillo M et al (2010) Paracrine signaling by glial cell-derived triiodothyronine activates neuronal gene expression in the rodent brain and human cells. J Clin Invest 120:2206–2217

Galton VA (2005) The roles of the iodothyronine deiodinases in mammalian development. Thyroid 15:823–834

Galton VA, Wood ET, St. Germain EA et al (2007) Thyroid hormone homeostasis and action in the type 2 deiodinase-deficient rodent brain during development. Endocrinology 148:3080–3088

Gereben B, Zavacki AM, Ribich S et al (2008) Cellular and molecular basis of deiodinase-regulated thyroid hormone signaling. Endocr Rev 29:898–938

Glinoer D (2001) Pregnancy and iodine. Thyroid 11:471–481

Guadano-Ferraz A, Obregon MJ, St. Germain DL et al (1997) The type 2 iodothyronine deiodinase is expressed primarily in glial cells in the neonatal rat brain. Proc Natl Acad Sci U S A 94:10391–10396

Hernandez A, Martinez ME, Fiering S et al (2006) Type 3 deiodinase is critical for the maturation and function of the thyroid axis. J Clin Invest 116:476–484

Hernandez A, Quignodon L, Martinez ME et al (2010) Type 3 deiodinase deficiency causes spatial and temporal alterations in brain T3 signaling that are dissociated from serum thyroid hormone levels. Endocrinology 151:5550–5558

Hernandez A, Morte B, Belinchón MM et al (2012) Critical role of type 2 and 3 deiodinases in the negative regulation of gene expression by T3 in the mouse cerebral cortex. Endocrinology 153:2919–2928

Horn S, Heuer H (2010) Thyroid hormone action during brain development: more questions than answers. Mol Cell Endocrinol 315:19–26

Hosoi Y, Murakami M, Mizuma H et al (1999) Expression and regulation of type 2 iodothyronine deiodinase in cultured human skeletal muscle cells. J Clin Endocrinol Metab 84:3293–3300

Kester MH, Martines de Mensa R, Obregon MJ et al (2004) Iodothyronine levels in the human developing brain: major regulatory roles of iodothyronine deiodinases in different areas. J Clin Endocrinol Metab 89:3117–3128

Koibuchi N, Jingu H, Iwasaki T et al (2003) Current perspectives on the role of thyroid hormone in growth and development of cerebellum. Cerebellum 2:279–289

LaFranchi SH, Austin J (2007) How should we be treating children with congenital hypothyroidism? J Pediatr Endocrinol Metab 20:559–578

Lavado-Autric R, Auso E, Garcia-Velasco JV et al (2003) Early maternal hypothyroxinemia alters histogenesis and cerebral cortex cytoarchitecture of the progeny. J Clin Invest 111:1073–1082

Mizuma H, Murakami M, Mori M (2001) Thyroid hormone activation in human vascular smooth muscle cells: expression of type 2 iodothyronine deiodinase. Circ Res 88:313–318

Morreale de Escobar G, Obregon MJ, Escobar Del Rey F (2004) Role of thyroid hormone during early brain development. Eur J Endocrinol 151(Suppl 3):U25–U37

Morte B, Bernal J (2014) Thyroid hormone action: astrocyte-neuron communication. Front Endocrinol 5:82. doi:10.3389/fendo.2014.00082

Murakami M, Araki O, Morimura T et al (2000) Expression of type 2 iodothyronine deiodinase in brain tumors. J Clin Endocrinol Metab 85:4403–4406

Murakami M, Araki O, Hosoi Y et al (2001) Expression and regulation of type 2 iodothyronine deiodinase in human thyroid gland. Endocrinology 142:2961–2967

Ng L, Goodyear RJ, Woods CA et al (2004) Hearing loss and retarded cochlear development in mice lacking type 2 iodothyronine deiodinase. Proc Natl Acad Sci U S A 101:3474–3479

Ng L, Hernandez A, He W et al (2009) A protective role for type 3 deiodinase, a thyroid hormone-inactivating enzyme, in cochlear development and auditory function. Endocrinology 150:1952–1960

Ng L, Lyubarsky A, Nikonov SS et al (2010) Type 3 deiodinase, a thyroid-hormone-inactivating enzyme, controls survival and maturation of cone photoreceptors. J Neurosci 30:3347–3357

Ng L, Kelley MW, Forrest D (2013) Making sense with thyroid hormone-the role of T3 in auditory development. Nat Rev Endocrinol 9:296–307

Obregon MJ, Calvo RM, Escobar del Rey F et al (2007) Ontogenesis of thyroid function and interactions with maternal function. Endocr Dev 10:86–98

Patel J, Landers K, Li H et al (2011) Thyroid hormones and fetal neurological development. J Endocrinol 209:1–8

Pearce EN (2009) What do we know about iodine supplementation in pregnancy? J Clin Endocrinol Metab 94:3188–3190

Peeters RP, Hernandez A, Ng L et al (2013) Cerebellar abnormalities in mice lacking type 3 deiodinase and partial reversal of phenotype by deletion of thyroid hormone receptor α1. Endocrinology 154:550–561

Porterfield SP, Hendrich CE (1993) The role of thyroid hormones in prenatal and neonatal neurological development-current perspectives. Endocr Rev 14:94–106

Schneider MJ, Fiering SN, Pallud SE et al (2001) Targeted disruption of the type 2 selenodeiodinase gene (DIO2) results in a phenotype of pituitary resistance to T4. Mol Endocrinol 15:2137–2148

Schroeder AC, Privalsky ML (2014) Thyroid hormones, T3 and T4, in the brain. Front Endocrinol 5:40. doi:10.3389/fendo.2014.00040

St. Germain DL, Galton VA, Hernandez A (2009) Minireview: defining the roles of the iodothyronine deiodinases: current concepts and challenges. Endocrinology 150(3):1097–1107

Tu HM, Kim SW, Salvatore D et al (1997) Regional distribution of type 2 thyroxine deiodinase messenger ribonucleic acid in rat hypothalamus and pituitary and its regulation by thyroid hormone. Endocrinology 138:3359–3368

Williams GR (2008) Neurodevelopmental and neurophysiological actions of thyroid hormone. J Neuroendocrinol 20:784–794

Chapter 3
Brominated Organohalogens and Neurodevelopment: Different Mechanisms, Same Consequence

Kingsley Ibhazehiebo, Toshiharu Iwasaki, and Noriyuki Koibuchi

Abstract Brominated organohalogens including polybrominated diphenyl ethers (PBDEs), hexabromocyclododecane (HBCD), tetrabromobisphenol A (TBBPA), and polybrominated biphenyl mixture (BP-6) are ubiquitous industrial chemicals used extensively as flame retardants in a wide range of consumer and household items including furnitures and electronics. Their ability to persist in the environment and bioaccumulate in humans and wildlife is currently of great health concern. These brominated organohalogens have been implicated in developmental neurotoxicity in numerous in vitro and in vivo models, acting through multiple mechanisms. In this chapter, we will examine different pathways and mechanisms of brominated organohalogen actions and their consequences for neurodevelopment. Although many mechanisms and pathways have been postulated for brominated organohalogen actions, they all converge at same consequences of impaired brain development.

Keywords Endocrine disrupting chemicals • Brain development • Thyroid hormone receptor • Transcription • Cerebellum

3.1 Introduction

Persistent organic pollutants (POPs) like brominated flame retardants (BFR) which include polybrominated diphenyl ethers (PBDEs) were introduced as the main flame retardants about five decades ago following the ban on another previously used organohalogen, polychlorinated biphenyls (PCBs) due to questions on their harmful effects on the environment and humans (Carpenter 2006; reviewed by Fonnum and Mariussen 2009; Seegal 1996). BFRs are a class of industrial chemicals showing

K. Ibhazehiebo • T. Iwasaki • N. Koibuchi (✉)
Department of Integrative Physiology, Gunma University Graduate School of Medicine,
3-39-22 Showa-machi, Maebashi, Gunma 371-8511, Japan
e-mail: nkoibuch@gunma-u.ac.jp

© Springer Science+Business Media New York 2016 33
N. Koibuchi, P.M. Yen (eds.), *Thyroid Hormone Disruption and Neurodevelopment*,
Contemporary Clinical Neuroscience, DOI 10.1007/978-1-4939-3737-0_3

structural diversity, and PBDE is a group of compounds containing 209 congeners with varying degree of bromination ranging from 1 to 9 bromine molecules (Costa et al. 2008; Dingemans et al. 2011). Commercially, PBDEs are produced as penta-bromodiphenyl ether, octabromodiphenyl ether, and decabromodiphenyl ether (Costa et al. 2008). Currently, it is estimated that the amount of PBDE produced in the last three decades is well over two million tons (Shaw and Kannan 2009). PBDEs are used primarily as flame retardants in a wide range of consumer and household goods including electrical and electronic products, vehicles, textiles, construction materials, and infant products to limit the damaging effects of fire outbreak (Darnerud et al. 2001; de Boer et al. 1998; Hale et al. 2001; IPCS 1994).

PBDEs are lipophilic, semi-volatile, and additive flame retardant (not covalently bound to the polymer), which makes them leach easily from products into the environment. Currently, PBDEs are regarded as a ubiquitous environmental contaminant with detectable levels found in birds, fish, marine mammals, dust sediments, and outdoor and indoor air (Costa and Giordano 2011; deWit 2002; Hites et al. 2004; reviewed by Law et al. 2003, 2008; Stapleton et al. 2005). This is due to widespread use and poor disposal over the last 30 years (Costa et al. 2008; Hites 2004). Human exposure to PBDE occurs via inhalation and ingestion of house dust especially among toddlers and via consumption of PBDE-contaminated plant and animal products (Frederiksen et al. 2009; Johnson-Restrepo and Kannan 2009). Recent studies suggest that levels of PBDE found in humans and environment are one order of magnitude higher in North America compared to the European Union and Asia (reviewed by Frederiksen et al. 2009). Detectable levels has been reported in breast milk, blood, fatty tissues in humans, and the liver of fetuses and newborns (Inoue et al. 2006; Morland et al. 2005; Schecter et al. 2005, 2007; She et al. 2002), and epidemiological studies suggest rapidly rising levels of PBDEs in human bodies over the past few decades (Costa et al. 2008). PBDEs are extremely stable environmental pollutant, resistant to biodegradation and in humans have a half-life ranging from about 2 to 12 years (Geyer et al. 2004). Extensive studies by Costa and Giordano (2007), Kierkegaard et al. (1999), Mc Donald (2005), Meironyte et al. (1999), and Sellstrom et al. (1993) identified 2,2′,4,4′-tetraBDE (BDE-47), 2,2′,4,4′,5-pentaBDE (BDE99), 2,2′,4,4′,6-pentaBDE (BDE100), 2,2′,4,4′,5,5′,6-hexaBDE (BDE153),2,2′,4,4′,5,6-hexaBDE (BDE154),and 2,2′,3,3′,4,4′,5,5′,6,6′-decaBDE (BDE209) as the predominant congeners found in human tissues accounting for 90 % of total body burden. DE71 and DE79 are the major PBDE mixtures (Costa and Giordano 2007), while decabrominated diphenyl ether (BDE 209) is the second most used brominated flame retardant (BFR) (Tseng et al. 2008). PBDE can cross the placenta as it has been detected in the fetal blood and liver (Gomera et al. 2007; Main et al. 2007), and BDE 209 is the predominant congener found in breast milk and the placenta (Gomera et al. 2007). Although production of penta- and octa-PBDEs has been halted in Europe and North America, human and environmental exposure continues from existing products containing these (Li et al. 2013). The fully brominated PBDE congener (BDE 209) is still produced and widely used in many countries (Hale et al. 2006; Shaw and Kannan 2009).

3.2 Thyroid Hormone Signaling and PBDE

Thyroid hormones (TH) triiodothyronine (T3) and thyroxine (T4) are essential for normal brain development in animals and humans. They play critical roles in many physiological events as well as key metabolic pathways (Harvey and Williams 2002). TH production begins about 10 weeks' gestation in humans, and levels rise continuously throughout gestation (Porterfield and Hendrich 1993; Porterfield 2000). TH homeostasis is regulated by a sensitive feedback loop within the hypothalamic-pituitary-thyroid (HPT) axis (Capen 1997; Gilbert et al. 2012). Thyroid-stimulating hormone (TSH) induces the thyroid to synthesize T4, which can rapidly cross the blood-brain barrier (BBB) and enter into the brain where it is de-iodinated by 5′-deiodinase enzyme contained in astrocytes to T3. T3 which is the more biologically active form is taken up by the neuronal cells and bound to TH receptor (Koibuchi et al. 2008). TH functions are biologically regulated by TRs. TRs are ligand-dependant transcription factors that are widely expressed in similar patterns (Bradley et al. 1992). TR is bound to specific DNA sequence, TH response element (TRE), composed of two half-sites core motifs (AGGTCA), with specific nucleotide spacing and orientation, upstream of the target gene. These TR-TRE complex interacts with the orphan receptor retinoid X receptors (RXR) to form heterodimers, which in turn interact with host of nuclear coactivators or repressors to mediate or silence transcription (Koibuchi et al. 1999, 2008; Takeshita et al, 1998, 2002).

THs control neuronal and glial proliferation in definitive brain regions and mediate neuronal migration and differentiation during the fetal and neonatal developmental windows (Porterfield 2000). This is achieved by the spatial and temporal regulation of genes involved in a variety of developmental processes (Gilbert et al. 2012). TH actions are essential during the entire period of brain development in humans (which extend into early childhood), but are more critical during the perinatal period of brain development, a phase commonly referred to as the "brain growth spurt" which extends from the third trimester of pregnancy to around second year after birth (Dobbing and Sands 1979). Altered maternal TH homeostasis preceding the onset of fetal TH activity could have grave consequences because the fetus is dependent on maternal TH for normal brain development (Chevrier 2013; Morreale de Escobar et al. 2000). Deficiency of TH (hypothyroidism) especially during the perinatal period causes abnormal brain development (cretinism) with severe cognitive and/or mental disorders in the offsprings including impaired attention, expressive language (Henrichs et al. 2010; Koibuchi and Chin 2000; Oppenheimer and Schwartz 1997; Yen 2001), as well as abnormal behavioral patterns (Haddow et al. 1999). The cerebellum is critical for motor learning, memory, and vestibular functions (Hirai 2008), and Purkinje cells are a cornerstone of higher neuronal function in the cerebellum being the sole output from the cerebellar cortex (Hirai 2008; Ito 2002). TH plays critical roles in early brain development especially in the cerebellum (Williams 2008). Therefore, neurodevelopmental disruption of cerebellar circuitry due to altered TH homeostasis could have grave implications. Specifically, hypothyroidism has been shown in the cerebellum to cause reduced cell numbers, reduced growth, and branching of Purkinje cell dendritic arborization, reduced synaptogenesis between the Purkinje cells and granule cell axons, delayed myelination, delayed proliferation and altered migration of granule cells, and changes in synaptic connection among cerebellar neurons and afferent neuronal fibers (Brown et al. 1976; Koibuchi and Chin 2000; Legrand 1980; Neveu and Arenas 1996; Nicholson and Altman 1972a, b, c; Williams 2008). The structural similarity of PBDE especially the OH-PBDE metabolites to TH is of great concern because THs are also hydroxyl-halogenated diphenyls and PBDE may act through TRs (Schriks et al. 2007) and disrupt TH homeostasis (Brouwer et al. 1998; Birbaum and Staskal 2004; Boas et al. 2006; Hooper and Mc Donald 2000; Stapleton et al. 2011). As PBDE can be transferred across the blood-brain barrier and accumulate in the central nervous system, it may act directly on TR in brain tissues to induce abnormal brain development (Naert et al. 2007).

3.3 Neurobehavioral Effects of PBDE

PBDEs are global environmental contaminants of great health concern (Hale et al. 2003; Law et al. 2003; Norstrom et al. 2002). Over the last decade, PBDE's ability to negatively impact neurobehavioral development has been extensively studied. Much of the investigations have focused on neonatal exposure to PBDE at specific developmental time points using rodent models. PBDE-exposed neonatal rodents showed aberrant motor and cognitive functions and performed poorly in learning tasks and other behavioral endpoints (Branchi et al. 2003; Mc Donald 2005). Also, altered spontaneous motor behavior and impaired learning and memory as well as persistent changes in the cholinergic system induced during neonatal exposure to PBDE have been reported in animal models (Eriksson et al. 2001; Viberg et al. 2003). In their study, mice were exposed to single oral dose (8 mg/kg) of lower brominated congeners (BDE 99) 10 days postnatally, and significant impairment in locomotor, rearing, learning, and total activities were observed 2 and 4 months later. Neonatal exposure to higher brominated PBDEs including BDE 203, 206, and 209 at doses ranging from 0.4 to 20 mg/kg has also been associated with developmental neurotoxicity by same authors (Viberg et al. 2002, 2006, 2008). Also, the same authors reported permanent abnormal changes in behavior following neonatal exposure to single oral doses of higher brominated BDEs in mice. Other investigators have also reported aberrant neurobehavioral effects in rodents exposed to PBDE during gestation and lactation periods of development (Gee and Moser 2008; Koenig et al. 2012; Rice et al. 2009). In a recent study, chronic perinatal exposure to BDE 47 was reported to alter development including somatic growth, cognition, and motor function (Ta et al. 2011). In humans, there is established correlation between maternal PBDE levels and neurodevelopment abnormalities in offsprings (Herbstman et al. 2010). PBDE exposure (pre- or postnatal) has been implicated in developmental neurotoxicity in children (Costa and Giordano 2007; Dingemans et al. 2011; Verner et al. 2011) and behavioral changes in adults including lower verbal learning and memory (Fitzgerald et al. 2012). Recent studies suggest that PBDE exposure may be a risk factor for autism (Mitchell et al. 2012; Napoli et al. 2013; Woods et al. 2012).

3.4 Mechanism of PBDE Action

3.4.1 In Vivo Evidence of PBDE Action

Rats exposed to BDE 209 have been shown to have increased incidence of thyroid adenomas and hepatocellular carcinomas (Darnerud et al. 2001; NTP 1986), while decreased serum levels of T4 in male mouse after exposure to BDE 209 have also been reported (Rice et al. 2007). Decreased levels of serum T4 with elevated cytochrome P450 activities have been shown in rats exposed to BDE 99 and DE-71 (PBDE mixture containing mainly tetra and pentaBDE) (Kuriyama et al. 2007; Zhou et al. 2002). Exposure of rats and mice to the PBDE mixture Bromkal 70-5DE, BDE-47, and DE-71 resulted in decreased levels of T4 coupled with induction of phase 1 (ethoxyresorufin-O-deethylase, pentoxyresorufin-O-deethylase) and phase 2 (uridine 5-diphosphateglucoronosyltransferase) metabolizing enzymes activities, indicating that the reduction in T4 levels may be due partially to increased conjugation of TH and subsequent biliary excretion (Fowles et al. 1994; Hallgren and Darnerud 1998; Skarman et al. 2005; Zhou et al. 2001). PBDE exposure in rodent models has been shown to disrupt TH homeostasis via inhibition of deiodinase 1 enzyme activity (Szabo et al. 2009). Also, PBDE mixture (DE 71) has been reported to significantly reduce levels of circulating TH and disrupt TH homeostasis by altering active transport, glucuronidation, sulfonation, and deiodination in rodent models (Kodavanti et al. 2010; Szabo et al. 2009). Lower brominated PBDE (BDE 99) was recently reported to evoke cytotoxicity in the cerebellum via oxidative stress in rodent models (Belles et al. 2010). PBDE has also been reported to alter levels of protein implicated in neurodegeneration and neuroplasticity in the striatum as well as disrupt energy metabolism in the hippocampus (Alm et al. 2006; Ta et al. 2011).

3.4.2 In Vitro Evidence of PBDE Action

Most of the evidence supporting PBDE disruption of TH signaling are from a variety of in vitro studies emanating from a host of laboratories worldwide.

Oxidative Stress: In several studies, both lower brominated BDEs and the fully brominated BDE were shown to cause cell death in both neuronal cell lines and primary cultured neurons (reviewed in Dingemans et al. 2011). This is due to increased oxidative stress as a result of reduced capacity of antioxidant enzymes to inactivate reactive oxygen species (ROS) following exposure to PBDEs, thereby causing oxidative damage to mitochondrial proteins, membranes, and DNA (reviewed in Dingemans et al. 2011; Murphy 2009).

Inhibition of Cellular Differentiation and Migration: PBDE exposure has also been reported to inhibit cellular differentiation and migration in several in vitro studies. Specifically, the fully brominated PBDE congener, BDE209, has been shown to inhibit neurite outgrowth and remarkably altered the differentiation of neural stem cells at very low dose (10 µM) (Zhang et al. 2010). The lower brominated congener BDE 99 has been reported to directly inhibit neurodifferentiation in PC12 cells, a well characterized model for neurodifferentiation (Slotkin et al. 2013). Also, human neural progenitor cells when exposed to 1 µM of the lower brominated PBDE congeners (BDE47 and BDE99) showed significantly reduced ability to differentiate into neurons or oligodendrocytes (Schreiber et al. 2010).

Impaired Cellular Transport: Further in vitro studies indicate that PBDE metabolites (hydroxylated PBDE; OH-PBDE) bind with high affinity to TH transport protein including transthyretin (TTR) and thyroxine-binding globulin (TBG) (Cao et al. 2010; Marchesini et al. 2008; Meerts et al. 2000), which are important transport proteins essential for the transport of T4 through the blood and into developing tissues like the fetal brain (McKinnon et al. 2005; Richardson et al. 2008). This action both competitively prevents T4 from binding with TTR leading to displacement and increased glucuronidation of T4 with consequent lower circulating levels (Costa and Giordano 2007; Hamers et al. 2006; Meerts et al. 2000) and also facilitates the accumulation of OH-PBDE in the developing fetus (Marchesini et al. 2008). However, thyroid hormone levels in the brain are not affected in TTR knockout mice suggesting that other pathways may be involved in thyroid hormone distribution in the brain (Palha et al. 2002).

Disruption of TH Signaling: Hydroxylated PBDEs have been shown to bind human thyroid hormone receptors in vitro (Kitamura et al. 2008; Marsh et al. 1998). Recent in vitro studies indicate that PBDEs and its hydroxylated metabolites can inhibit T3 activation of TR in reporter gene assays (Kojima et al. 2009; Schriks et al. 2007) or act as TH agonist (Hamers et al. 2006). The binding affinity of OH-PBDEs to TR could depend on the degree of bromination (Ren et al. 2013). Also, BDE 99 has been shown to directly decrease TR alpha 1 and 2 gene expression and also disrupt TR-regulated T3-responsive genes (Blanco et al. 2011). These data suggest that PBDE may affect various TH-regulated neuronal pathways essential for normal brain development and functions.

More recently, a variety of PBDE congeners including pentaBDE (BDE 100), hexaBDE (BDE 153, BDE 154), decaBDE (BDE 209), and PBDE mixture (DE 71) has been shown to disrupt TH signaling pathway in vitro in similar manner as PCBs (Ibhazehiebo et al. 2011a; Iwasaki et al. 2002, 2008; Miyazaki et al. 2004, 2008). Specifically, PBDE showed congener-specific suppression of TR-mediated transcription in fibroblast-derived CV-1 cells using the transient transfection-based reporter gene assay (Fig. 3.1). While the lower brominated congeners including BDE 28, BDE 47, BDE 66, and BDE 99 did not show any suppression of TR-mediated transcription, significant suppression of transcription (45 %) was observed with BDE 100 and BDE 209 at dose as low as 10 pM (Ibhazehiebo et al. 2011a). The reason for suppressed transcription was due to PBDE's ability to partially dissociate TR-TRE binding in vitro using the liquid chemiluminescence DNA

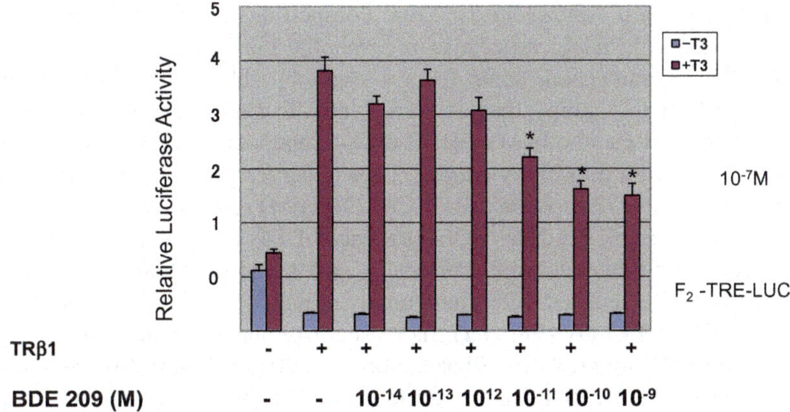

Fig. 3.1 PBDE showed congener-specific suppression of TR-mediated transcription in fibroblast-derived CV-1 cells using the transient transfection-based reporter gene assay

pull-down assay. Specifically, BDE 209 (1 nM) caused 40 % partial dissociation of TR-TRE complex in the presence of 1 µM T3. Also, PBDE has been shown to disrupt TH signaling acting through the DNA-binding domain (DBD) of TR (Ibhazehiebo et al. 2011a). In their study, the authors reported that all chimeras harboring TR-DBD showed suppression of transcription by BDE 209 regardless of the difference in ligand binding domain, such differences were not seen with chimeras harboring GR-DBD, suggesting that the site of action of PBDE may be DBD of TR (Ibhazehiebo et al. 2011a). Interestingly, other brominated cyclic alkanes like 1,2,5,6,9,10-alpha-hexabromocyclododecane (HBCD) and the polybrominated biphenyl mixture BP-6 have been shown to similarly suppress TR-mediated transcription (Ibhazehiebo et al. 2011b, c).

Altered Neuronal and Calcium Ion Signaling Pathways: PBDE and its hydroxyl metabolites have been reported to also alter neuronal signaling via inhibition of synaptosomal and vesicular neurotransmitter uptake, induce vesicular neurotransmitter release by exocytosis, and reduce synaptosomal dopamine levels by inhibiting the activities of the membrane dopamine transporter (Dingemans et al. 2007; Dreiem et al. 2010; Fonnum and Mariussen 2009). There is also abundant evidence that PBDE significantly altered several aspects of calcium ion intracellular signaling pathways by enhancing protein kinase C translocation (Dingemans et al. 2011; Kodavanti and Derr-Yellin 2002; Kodavanti and Ward 2005; Londoño et al. 2010; reviewed in Dingemans et al. 2011) and promoting the rapid release of arachidonic acid following activation of phospholipase A (Madia et al. 2004; reviewed in Dingemans et al. 2011). Also, PBDE is implicated in the activation of the mitogen-activated protein kinase (MAP kinase) pathway which is essential for various intracellular activities (Fan et al. 2010; reviewed in Dingemans et al. 2011; Pearson et al. 2001). PBDE has also been reported to act directly on the developing brain by altering levels of cholinergic receptor in the hippocampus (Eriksson et al. 2002; Viberg et al. 2003), altering CaMKll, BDNF, and GAP-43 levels (Viberg et al. 2003, 2008) and reducing long-term potentiation (LTP) in the hippocampus (Dingemans et al. 2007).

Altered Cerebellar Neurogenesis: More compelling in vitro evidences for the neurotoxic effects of PBDEs especially on cerebellar Purkinje cell and its presynaptic partner, the granule cell, came from a study by Ibhazehiebo and coworkers (2011a, b, c, d). In their study, they examined the effect of several PBDE congeners as well as hydroxyl metabolites of PBDE on T4-dependent cerebellar Purkinje cell dendritic arborization in primary culture. They observed that 17 days after onset of culture, low-dose PBDE especially BDE 209 (100 pM) remarkably inhibited development of Purkinje cell dendrites in the presence of T4. The dendrites showed poor growth, secondary branches were remarkably small, and quantitatively, the area of these Purkinje cell dendrites were dramatically reduced compared to those cultured in the presence of T4 only (Fig. 3.2). Increased T4 concentrations could only partially relieve PBDE suppression (Ibhazehiebo et al. 2011a). They also observed that in the absence of T4, PBDE-treated Purkinje cells exhibited an almost complete absence of dendrites. These effects by PBDE were however congener-specific, as not all PBDE tested inhibited Purkinje cell dendritic development (Ibhazehiebo et al. 2011a). Similar suppression of Purkinje cell dendritic outgrowth has also been observed with brominated cyclic alkane like HBCD and PBDE mixture BP-6 (Ibhazehiebo et al. 2011b, c). Cerebellar Purkinje cells are a good model to examine various TH functions and also to study consequences of TH system disruption because dendritic growth of Purkinje cells provides a good index of their development and function (Kimura-Kuroda et al. 2002, 2007). As TH tightly regulates fundamental gene expression both directly and indirectly in vast brain regions including the cerebellum (Koibuchi et al. 2001), the inhibitory effects of PBDE and other brominated alkanes like HBCD on TR-mediated gene expression could widely disrupt normal brain development acting via TH-dependent gene regulation (Ibhazehiebo et al. 2011a). Furthermore, HBCD, PBDE, and PBDE mixture (BP-6) has been shown to dramatically reduce in real-time TH-induced neurite extension of cerebellar granule cell at low dose (Ibhazehiebo et al. 2011b, d; Ibhazehiebo et al., manuscript in preparation). HBCD was observed to reduce cerebellar granule cell neurite extension at 100 pM concentration (Fig. 3.3), and increased TH could not relieve HBCD suppression of neurite outgrowth (Ibhazehiebo et al. 2011d). A recent study suggests that hydroxylated metabolites of PBDE-47 can interfere with adult neurogenesis by inhibition of oligodendrocytes and neurons differentiation, inhibition of cell proliferation, interference with MAP kinase signaling pathway, and disruption of brain neurotransmitter (NT3) function (Li et al. 2013). Altered adult neurogenesis could adversely perturb normal adult brain function.

3.5 Summary

Although the production and use of most brominated flame retardants including PBDE have been greatly reduced, environmental and human exposures still persist due to the presence of many consumer and industrial products containing these flame retardants and their persistence in the environment. While BFRs act via several mechanisms and pathways in the brain, the ultimate consequence has been

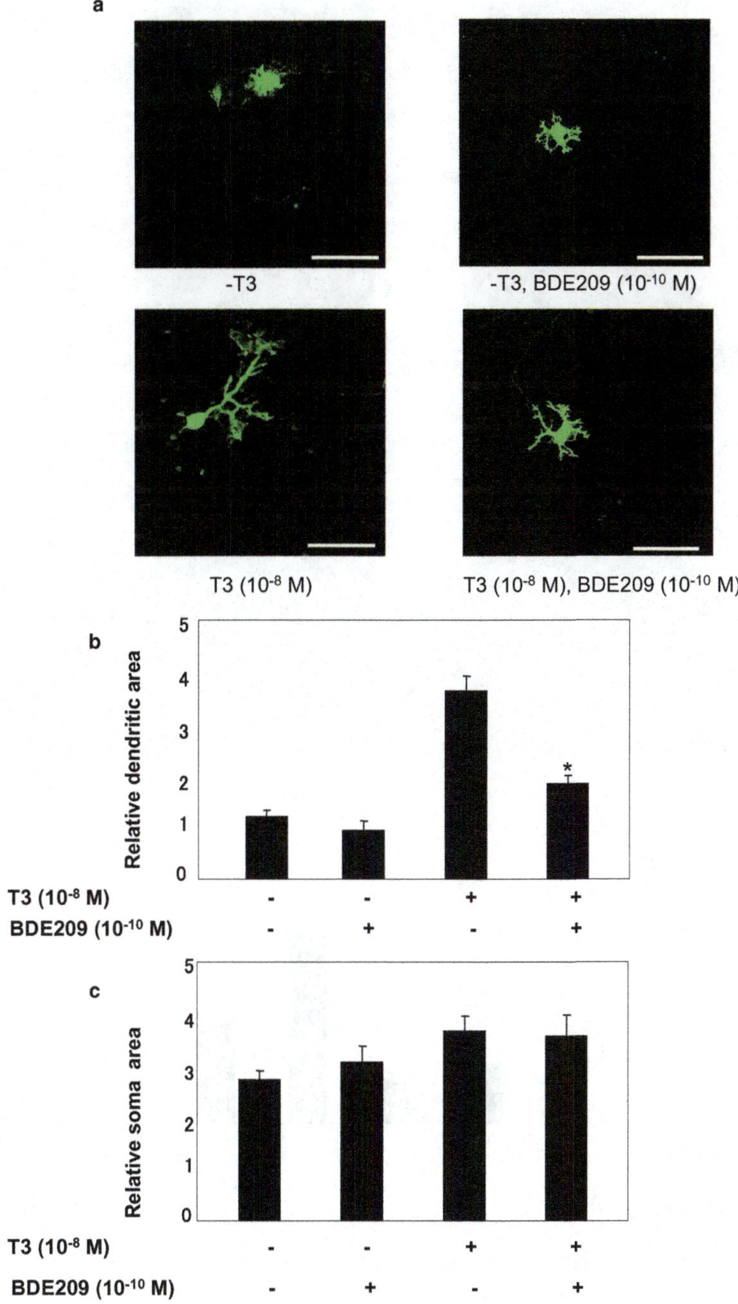

Fig. 3.2 Low-dose PBDE especially BDE 209 (100 pM) remarkably inhibited development of Purkinje cell dendrites in the presence of T4. The dendrites showed poor growth, and secondary branches were remarkably small, and quantitatively, the area of these Purkinje cell dendrites were dramatically reduced compared to those cultured in the presence of T4 only

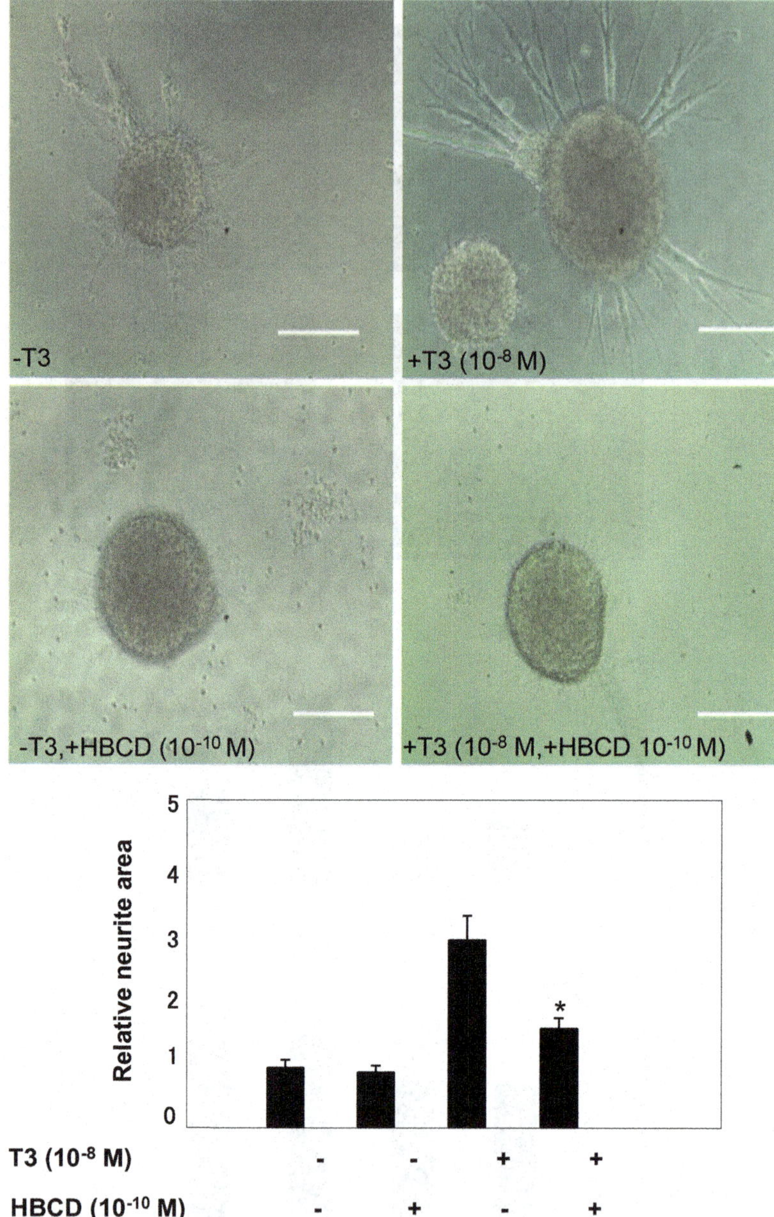

Fig. 3.3 HBCD observed to reduce cerebellar granule cell neurite extension at 100 pM concentration

shown to be impaired brain development. More knowledge is however needed to further elucidate and clarify mechanisms of PBDE action, consequences of co-exposure of various BFRs, and the sensitivity to these BFRs during specific periods of brain development.

References

Alm H, Scholz B, Fischer C, Kultima K, Viberg H, Eriksson P, Dencker L, Stigson M (2006) Proteomic evaluation of neonatal exposure to 2,2,4,4,5-pentabromodiphenyl ether. Environ Health Perspect 114:242–259

Belles M, Alonso V, Linares V, Albina ML, Sirvent JJ, Domingo JL et al (2010) Behavioral effects and oxidative status in brain regions of adult rats exposed to BDE-99. Toxicol Lett 194:1–7

Birbaum LS, Staskal DF (2004) Brominated flame retardant: cause for concern? Environ Health Perspect 112:9–12

Blanco J, Mulero M, Lopez M, Domingo JL, Sanchez DJ (2011) BDE-99 deregulates BDNF, Bcl-2 and the mRNA expression of thyroid receptor isoforms in rat cerebellar granular neurons. Toxicology 290:305–311

Boas M, Feldt-Rasmussen U, Skakkebaek NE, Main KM (2006) Environmental chemicals and thyroid function. Eur J Endocrinol 154:599–611

Bradley DJ, Towle HC, Young WS (1992) Spatial and temporal expression of α- and β-thyroid hormone receptor mRNAs, including the β2 subtype, in the developing mammalian nervous system. J Neurosci 12:2288–2302

Branchi IF, Capone F, Alleva E, Costa LG (2003) Polybrominated diphenyl ethers: neurobehavioral effects following developmental exposure. Neurotoxicology 24:449–462

Brouwer A, Morse DC, Lans MC, Schuur AG, Murk AJ, Klasson-Wehler E et al (1998) Interactions of persistent environmental organohalogens with the thyroid hormone system: mechanisms and possible consequences for animal and human health. Toxicol Ind Health 14:59–84

Brown WJ, Verity MA, Smith RL (1976) Inhibition of cerebellar dendritic development in neonatal thyroid deficiency. Neuropathol Appl Neurobiol 2:191–207

Cao J, Lin Y, Guo LH, Zhang AQ, Wei Y, Yang Y (2010) Structure-based investigation on the binding interaction of hydroxylated polybrominated diphenyl ethers with thyroxine transport proteins. Toxicology 277:20–28

Capen CC (1997) Mechanistic data and risk assessment of selected toxic endpoints of the thyroid gland. Toxicol Pathol 25:39–48

Carpenter DO (2006) Polychlorinated biphenyl (PCBs); routes of exposure and effects on human health. Rev Environ Health 21:1–23

Chevrier J (2013) Maternal plasma polybrominated diphenyl ethers and thyroid hormone—challenges and opportunities. Am J Epidemiol 178(5):714–719. doi:10.1093/aje/kwt138

Costa LG, Giordano G (2007) Developmental neurotoxicity of polybrominated diphenyl ether (PBDE) flame retardants. Neurotoxicology 28:1047–1067

Costa LG, Giordano G (2011) Is decabromodiphenyl ether (BDE-209) a developmental neurotoxicant? Neurotoxicology 32:9–24

Costa LG, Giordano G, Tagliaferri S, Caglieri A, Mutti A (2008) Polybrominated diphenyl ether (PBDE) flame retardants: environmental contamination, human body burden and potential adverse health effects. Acta Biomed 79:172–183

Darnerud PO, Eriksen GS, Johannesson T, Larsen PB, Viluksela M (2001) Polybrominated diphenyl ethers: occurrence, dietary exposure and toxicology. Environ Health Perspect 109:49–68

de Boer J, Wester PG, Klamer HJ, Lewis WE, Boon IP (1998) Do flame retardants threaten ocean life? Nature 394:28–29

deWit CA (2002) An overview of brominated flame retardants in the environment. Chemosphere 46:583–624

Dingemans MM, Ranmakers GM, Gardoni F, van Kleef RG, Bergman A, Di Luca M et al (2007) Neonatal exposure to brominated flame retardant BDE-47 reduces long-term potentiation and postsynaptic protein levels in mouse hippocampus. Environ Health Perspect 115:865–870

Dingemans MML, van den Berg M, Westerink RHS (2011) Neurotoxicity of brominated flame retardants: (in)direct effects of parent and hydroxylated polybrominated diphenyl ethers on the (developing) nervous system. Environ Health Perspect 119:900–907

Dobbing J, Sands J (1979) Comparative aspects of the brain growth spurt. Early Hum Dev 3:79–83

Dreiem A, Okoniewski RJ, Brosch KO, Miller VM, Seegal RF (2010) Polychlorinated and polybrominated biphenyl ethers alter stratal dopamine neurochemistry synaptosomes from developing rats in an additive manner. Toxicol Sci 118:150–159

Eriksson P, Jakobsson E, Fredriksson A (2001) Brominated flame retardant: a novel class of developmental neurotoxicants in our environment? Environ Health Perspect 109:903–908

Eriksson P, Viberg H, Jakobsson E, Orb U, Fredriksson A (2002) A brominated flame retardant, 2,2′,4,4′,5-pentabromodiphenyl ether: uptake, retention, and induction of neurobehavioral alterations in mice during a critical phase of neonatal brain development. Toxicol Sci 67:98–103

Fan CY, Besas J, Kodavanti PR (2010) Changes in mitogen-activated kinase in cerebellar granule neurons by polybrominated diphenyl ethers and polychlorinated biphenyls. Toxicol Appl Pharmacol 245:1–8

Fitzgerald EF, Shrestha S, Gomez MI, McCaffrey RJ, Zimmerman EA, Kannan K, Hwang SA (2012) Polybrominated biphenyl ethers (PBDEs), polychlorinated biphenyls (PCBs) and neuropsychological status among older adults in New York. Neurotoxicology 33:8–15

Fonnum F, Mariussen E (2009) Mechanism involved in the neurotoxic effect effects of environmental toxicants such as polychlorinated biphenyls and brominated flame retardants. J Neurochem 111:1327–1347

Fowles FR, Fairbrother A, Baecher-Steppan L, Kerkvliet NI (1994) Immunologic and endocrine effects of the flame retardant pentabromodiphenyl ether (DE-71) in C57/BL/6J mice. Toxicology 86:49–61

Frederiksen M, Vorkamp K, Thomsen M, Knudsen LE (2009) Human internal and external exposure to PBDEs- a review of levels and sources. Int J Hyg Environ Health 212:109–134

Gee JR, Moser VC (2008) Acute postnatal exposure to brominated diphenyl ether 47 delays neuromotor ontogeny and alters motor activity in mice. Neurotoxicol Teratol 30:79–87

Geyer HJ, Schramm K-W, Danerud PO, Aune M, Feicth A, Fried KW (2004) Terminal elimination half-lives of the brominated flame retardants TBBPA, HBCD, and lower brominated PBDEs in humans. Organohalogen Compd 66:3867–3872

Gilbert ME, Rovet J, Chen Z, Koibuchi N (2012) Developmental thyroid hormone disruption: prevalence, environmental contaminants and neurodevelopmental consequences. Neurotoxicology 33:842–852

Gomera B, Herero L, Ramos JJ, Mateo JR, Fernandez MA, Garcia JF et al (2007) Distribution of polybrominated diphenyl ethers in human umbilical cord serum, paternal serum, maternal serum, placentas, and breast milk from Madrid population, Spain. Environ Sci Technol 41:6961–6968

Haddow JE, Palomoki GE, Allan WC, Williams JR, Knight GJ, Gagnon J et al (1999) Maternal thyroid deficiency during pregnancy and subsequent development of the child. N Engl J Med 341:549–555

Hale RC, La Guardia MJ, Harvey EP, Gaylor MO, Mainor TM, Duff WH (2001) Flame retardants: persistent pollutants in land-applied sludges. Nature 412:140–141

Hale R, Alaee M, Manchester-Neesvig JB, Stapleton HM, Ikonomou MG (2003) Polybrominated diphenyl ether flame retardants in the North American environment. Environ Int 29:771–779

Hale RC, La Guardia MJ, Harvey E, Gaylor MO, Mainor TM (2006) Brominated flame retardant concentrations and trends in abiotic media. Chemosphere 64:181–186

Hallgren S, Darnerud PO (1998) Effects of polybrominated diphenyl ethers (PBDEs), polychlorinated biphenyls (PCBs) and chlorinated paraffins (CPs) on thyroid hormone levels and enzyme activities in rats. Organohalogen Compd 35:391–394

Hamers T, Kamstra JH, Sonnervald E, Murk AJ, Kester MHA, Andersson PL et al (2006) In vitro profiling of the endocrine-disrupting potency of brominated flame retardants. Toxicol Sci 92:157–173

Harvey CB, Williams GR (2002) Mechanism of thyroid hormone action. Thyroid 12:441–446

Henrichs J, Bongers-Schokkings JJ, Schenk JJ, Ghassabian A, Schmidt HG, Visser TJ et al (2010) Maternal thyroid function during early pregnancy and cognitive functioning in early childhood: the generation R study. J Clin Endocrinol Metab 95:4227–4234

Herbstman JB, Sjodin A, Kurzon M, Lederman SA, Jones RS, Rauh V et al (2010) Prenatal exposure to PBDEs and neurodevelopment. Environ Health Perspect 118:712–719

Hirai H (2008) Progress in transduction of cerebellar purkinje cells *in vivo* using viral vectors. Cerebellum 7(3):273–278

Hites RA (2004) Polybrominated diphenyl ethers in the environment and in people: a meta analysis of concentrations. Environ Sci Technol 38:945–956

Hites RA, Foran JA, Schwager SJ, Knuth BA, Hamilton MC, Carpenter DO (2004) Assessment of polybrominated diphenyl ether in farmed and wild salmon. Environ Sci Technol 38:4945–4949

Hooper K, Mc Donald TA (2000) The PBDEs: an emerging environmental challenge and another reason for breast- milk monitoring programme. Environ Health Perspect 108:387–392

Ibhazehiebo K, Iwasaki T, Kimura-Kuroda J, Miyazaki W, Shimokawa N, Koibuchi N (2011a) Disruption of thyroid hormone receptor-mediated transcription and thyroid hormone-induced Purkinje cell dendrite arborization by polybrominated biphenyl ethers. Environ Health Perspect 119:168–175

Ibhazehiebo K, Iwasaki T, Shimokawa N, Koibuchi N (2011b) 1,2,5,6,9,10-alphaHexabromocyclododecane (HBCD) impairs thyroid hormone-induced dendrite arborization of Purkinje cells and suppress thyroid hormone receptor-mediated transcription. Cerebellum 10:22–31

Ibhazehiebo K, Iwasaki T, Okano-Uchida T, Shimokawa N, Ishizaki Y, Koibuchi N (2011c) Suppression of thyroid hormone receptor-mediated transcription and disruption of thyroid hormone-induced cerebellar morphogenesis by the polybrominated biphenyl mixture, BP-6. Neurotoxicology 32:400–409

Ibhazehiebo K, Iwasaki T, Xu M, Shimokawa N, Koibuchi N (2011d) Brain-derived neurotrophic factor (BDNF) ameliorates the suppression of thyroid hormone-induced granule cell neurite extension by hexabromocyclododecane (HBCD). Neurosci Lett 493:1–7

Inoue K, Harada K, Takenaka K, Uehara S, Kono M, Shimizu T et al (2006) Levels and concentration ratios of polychlorinated biphenyls, organochlorine pesticides and polybrominated biphenyl ethers in serum and breast milk in Japanese mothers. Environ Health Perspect 114:1179–1185

IPCS (1994) Environmental health criteria 162: brominated diphenyl ethers. International programme on chemical safety. World Health Organization, Geneva

Ito M (2002) Historical review of the significance of the cerebellum and the role of Purkinje cells in motor learning. Ann N Y Acad Sci 978:273–288

Iwasaki T, Miyazaki W, Takeshita A, Kuroda Y, Koibuchi N (2002) Polychlorinated biphenyls suppress thyroid hormone-induced transactivation. Biochem Biophys Res Commun 299:384–388

Iwasaki T, Miyazaki W, Rokutanda N, Koibuchi N (2008) Liquid chemiluminescent DNA pull-down assay to measure nuclear receptor-DNA binding in solution. Biotechniques 45:445–448

Johnson-Restrepo B, Kannan K (2009) An assessment of sources and pathways of human exposure to polybrominated diphenyl ethers in the United States. Chemosphere 76:542–548

Kierkegaard A, Sellstrom U, Bignert A, Olsson M, Asplund L, Jansson B et al (1999) Temporal trends of a polybrominated diphenyl ether (PBDE), a methoxylated PBDE and hexabromocyclododecane (HBCD) in Swedish biota. Organohalogen Compd 40:367–370

Kimura-Kuroda J, Nagata I, Negishi-Kato M, Kuroda Y (2002) Thyroid hormone-dependent development of mouse cerebellar Purkinje cells *in vitro*. Brain Res Dev Brain Res 137:55–65

Kimura-Kuroda J, Nagata I, Kuroda Y (2007) Disrupting effects of hydroxy-polychlorinated biphenyl (PCB) congeners on neuronal development of cerebellar Purkinje cells: a possible causal factor for developmental brain disorders? Chemosphere 67:412–420

Kitamura S, Shinohara S, Iwase E, Sugihara K, Uramaru N, Shigematsu H et al (2008) Affinity for thyroid hormone and estrogen receptors of hydroxylated polybrominated diphenyl ethers. J Health Sci 54:607–614

Kodavanti PR, Derr-Yellin EC (2002) Differential effects of polybrominated diphenyl ether and polychlorinated biphenyls on (^{3}H)arachidonic acid release in rat cerebellar granule neurons. Toxicol Sci 68:451–457

Kodavanti PR, Ward TR (2005) Differential effects of commercial polybrominated diphenyl ether and polychlorinated biphenyl mixtures on intracellular signaling in rat brain in vitro. Toxicol Sci 85:952–962

Kodavanti PR, Coburn CG, Moser VC, MacPhail RC, Fenton SE, Stoker TE et al (2010) Developmental exposure to a commercial PBDE mixture, DE-71: neurobehavioural, hormonal, and reproductive effects. Toxicol Sci 116:297–312

Koenig CM, Lango J, Pessah IN, Berman RF (2012) Maternal transfer of BDE-47 to offspring and neurobehavioral development in C57BL/6J mice. Neurotoxicol Teratol 36:571–580

Koibuchi N, Chin WW (2000) Thyroid hormone action and brain development. Trends Endocrinol Metab 11:123–128

Koibuchi N, Liu Y, Fukuda H, Takeshita A, Yen PM, Chin WW (1999) ROR alpha augments thyroid hormone receptor-mediated transcriptional activation. Endocrinology 140:1356–1364

Koibuchi N, Yamaoka S, Chin WW (2001) Effect of altered thyroid status on neurotrophin gene expression during postnatal development of the mouse cerebellum. Thyroid 11:205–210

Koibuchi N, Qiu CH, Miyazaki W, Iwasaki T, Shimokawa N (2008) The role of thyroid hormone in developing cerebellum. Cerebellum 7:499–500

Kojima H, Takeuchi S, Uramaru N, Sugihara K, Yoshida T, Kitamura S (2009) Nuclear hormone receptor activity of polybrominated diphenyl ether and their hydroxylated and methoxylated metabolites in transactivation assay using Chinese hamster ovary cells. Environ Health Perspect 117:1210–1218

Kuriyama SN, Wanner A, Fidalgo-Neto AA, Talsness CE, Koerner W, Chahoud I (2007) Developmental exposure to low-dose PBDE-99: tissue distribution and thyroid hormone levels. Toxicology 242:80–90

Law RJ, Alaee M, Allchin CR, Boon JP, Lebeuf M, Lepom P et al (2003) Levels and trends of polybrominated diphenyl ethers and other brominated flame retardants in wild life. Environ Int 29:757–770

Law RJ, Herzke D, Harrad S, Morris S, Bersuder P, Allchin CR (2008) Levels and trends of HBCD and BDEs in the European and Asian environments, with some information for other BFRs. Chemosphere 73:223–241

Legrand J (1980) Effects of thyroid hormone on brain development, with particular emphasis on glial cells and myelination. In: Di Benedetta C et al (eds) Multidisciplinary approach to brain development. Elsevier, Amsterdam, pp 279–292

Li T, Wang W, Pan Y-W, Xu L, Xia Z (2013) A hydroxylated metabolite of flame-retardant PBDE-47 decreases the survival, proliferation, and neuronal differentiation of primary cultured adult neuronal stem cells and interferes with signalling of ERK5 MAP kinase and neurotrophin 3. Toxicol Sci 134:111–124

Londoño M, Shimokawa N, Miyazaki W, Iwasaki T, Koibuchi N (2010) Hydroxylated PCB induces Ca2+ oscillations and alterations of membrane potential in cultured cortical cells. J Appl Toxicol 30:334–342

Madia F, Giordano G, Fattori V, Vitalone A, Branchi I, Capone F et al (2004) Differential in vitro neurotoxicity of the flame retardant PBDE-99 and of the PCB Aroclor 1254 in human astrocytoma cells. Toxicol Lett 154:11–21

Main KM, Kiviranta H, Virtanen HE, Sundqvist E, Tuomisto JT, Tuomisto J et al (2007) Flame retardants in placenta and breast milk and cryptorchidism in new born boys. Environ Health Perspect 115:1519–1526

Marchesini GR, Meimaridou A, Haasnoot W, Meulenberg E, Albertus F, Mizuguchi M, Takeuchi M, Irth H, Murk AJ (2008) Biosensor discovery of thyroxine transport disrupting chemicals. Toxicol Appl Pharmacol 252:150–160

Marsh G, Bergman A, Bladh LG, Gilner M, Jakobsson E (1998) Synthesis of p-hydroxybromodiphenyl ethers and binding to the thyroid receptor. Organohalogen Compd 37:305–308

Mc Donald TA (2005) Polybrominated diphenyl ether levels among United States residents: daily intake and risk of harm to developing brain and reproductive organs. Integr Environ Assess Manag 1:343–354

McKinnon B, Li H, Richard K, Mortimer R (2005) Synthesis of thyroid hormone binding proteins transthyretin and albumin by human trophoblast. J Clin Endocrinol Metab 90:6714–6720

Meerts IA, Van Zenden JJ, Luijks EA, Leeuwen-Bol I, Marsh G, Jakobsson EA et al (2000) Potent competitive interactions of some brominated flame retardants and related compounds with human transthyretin in vitro. Toxicol Sci 56:95–104

Meironyte D, Noren K, Bergman A (1999) Analysis of polybrominated diphenyl ethers in Swedish human milk: a time-related trend study, 1972–1997. J Toxicol Environ Health A58:329–341

Mitchell MM, Woods R, Chi LH, Schimidt RJ, Pessah IN, Kostyniak PJ, LaSalle JM (2012) Levels of select PCB and PBDE congeners in human postmortem brain reveal possible environmental involvement in 15q11-q13 duplication autism spectrum disorder. Environ Mol Mutagen 53:589–598

Miyazaki W, Iwasaki T, Takeshita A, Kuroda Y, Koibuchi N (2004) Polychlorinated biphenyls suppress thyroid hormone receptor-mediated transcription through a novel mechanism. J Biol Chem 279:18195–18202

Miyazaki W, Iwasaki T, Takeshita A, Tohyama C, Koibuchi N (2008) Identification of the functional domain of thyroid hormone receptor responsible for polychlorinated biphenyl-mediated suppression of its action in vitro. Environ Health Perspect 116:1231–1236

Morland KB, Landrigen PJ, Sjodin A, Gobeille AK, Jones RS, Mc Gahee EE et al (2005) Body burden of polybrominated diphenyl ethers among urban anglers. Environ Health Perspect 113:1689–1692

Morreale de Escobar G, Obregon MJ, Escobar del Rey F (2000) Is neuropsychological development related to maternal hypothyroidism or to maternal hypothyroxinemia? J Clin Endocrinol Metab 85:3975–3987

Murphy MP (2009) How mitochondria produce reactive oxygen species. Biochem J 417:1–13

Naert C, Van Peteqhem C, Kupper J, Jenni L, Naeqeli H (2007) Distribution of polychlorinated biphenyls and polybrominated diphenyl ethers in birds of prey from Switzerland. Chemosphere 68:977–987

Napoli E, Hung C, Wong S, Giulivi C (2013) Toxicity of the flame-retardant BDE-49 on brain mitochondria and neuronal progenitor stratal cells enhanced by a PTEN-deficient background. Toxicol Sci 132:196–210

National Toxicology programme (NTP) (1986) Toxicology and carcinogenesis studies of decabromodiphenyl oxides (CAS no 1163-19-5) in F344/N and B6C3F1 mice (Feed studies). Natl Toxicol Program Tech Rep Ser 309:1–242

Neveu I, Arenas E (1996) Neurotrophins promote the survival and development of neurons in the cerebellum of hypothyroid rats in vivo. J Cell Biol 133:631–646

Nicholson JL, Altman J (1972a) The effect of early hypo- and hyperthyroidism on the development of the rat cerebellar cortex. II. Synaptogenesis in the molecular layer. Brain Res 44:25–36

Nicholson JL, Altman J (1972b) Synaptogenesis in the rat cerebellum: effects of early hypo- and hyperthyroidism. Science 176:530–532

Nicholson JL, Altman J (1972c) The effect of early hypo- and hyperthyroidism on development of rat cerebellar cortex. I. Cell proliferation and differentiation. Brain Res 44:13–23

Norstrom RJ, Simon M, Moisey J, Wakeford B, Weseloh DVC (2002) Geographic distribution (2002) and temporal trends (1981-2000) of brominated diphenyl ethers in Great Lake herring gull eggs. Environ Sci Technol 36:4783–4789

Oppenheimer JH, Schwartz HL (1997) Molecular basis of thyroid hormone-dependent brain development. Endocr Rev 18:462–475

Palha JA, Nissanov J, Fernandes R, Sousa JC, Bertrand L, Dratman MB et al (2002) Thyroid hormone distribution in the mouse brain: the role of transthyretin. Neuroscience 113:837–847

Pearson G, Robinson F, Beers Gibson T, Xu BE, Karandikar M, Berman K et al (2001) Mitogen-activated protein (MAP) kinase pathways: regulations and physiological functions. Endocr Rev 22:153–183

Porterfield SP (2000) Thyroidal dysfunction and environmental chemicals—potential impact on brain development. Environ Health Perspect 108:433–438

Porterfield SP, Hendrich CE (1993) The role of thyroid hormones in prenatal and neonatal neurologic development: current perspectives. Endocr Rev 14:94–106

Ren XM, Guo LH, Gao Y, Zhang BT, Wan B (2013) Hydroxylated polybrominated biphenyl ethers exhibit different activities on thyroid hormone receptors depending on their degree of bromination. Toxicol Appl Pharmacol 268:256–263

Rice DC, Reeve EA, Herlihy A, Zweller RT, Thompson WD, Markowski VP (2007) Developmental delays and locomotor activity in the C57BL6/J mouse following neonatal exposure to the fully brominated PBDE, decabromodiphenyl ether. Neurotoxicol Teratol 29:511–520

Rice DC, Thompson WD, Reeve EA, Onos KD, Assahdollahzadeh M, Markowski VP (2009) Behavioural changes in aging but not young mice following neonatal exposure to the polybrominated flame retardant decaBDE. Environ Health Perspect 117:1903–1911

Richardson VM, Staskal DF, Ross DG, Diliberto JJ, DeVito MJ, Birnbaum LS (2008) Possible mechanisms of thyroid hormone disruption in mice by BDE 47, a major polybrominated diphenyl ether congener. Toxicol Appl Pharmacol 226:244–250

Schecter A, Papke O, Tung KC, Joseph J, Harris TR, Dahlgren J (2005) Polybrominated diphenyl ether flame retardants in the US population: current levels, temporal trends, and comparison with dioxins, dibenzofurans, and polychlorinated biphenyls. J Occup Environ Med 47:199–211

Schecter A, Johnson-Welch S, Tung KC, Harris TR, Papke O, Rosen R (2007) Polybrominated diphenyl ether (PBDE) levels in liver of U.S. human fetuses and newborns. J Toxicol Environ Health A 70:1–6

Schreiber T, Gassmann K, Gotz C, Hubenthal U, Moors M, Krause G et al (2010) Polybrominated biphenyl ethers induce developmental neurotoxicity in a human *in vitro* model: evidence for endocrine disruption. Environ Health Perspect 118:572–578

Schriks M, Roessig JM, Murk AJ, Furlow JD (2007) Thyroid hormone receptor isoform selectivity of thyroid hormone disrupting compounds quantified with an in vitro reporter gene assay. Environ Toxicol Pharmacol 23:302–307

Seegal RF (1996) Epidemiological and laboratory evidence of PCB-induced neurotoxicity. Crit Rev Toxicol 26:709–737

Sellstrom U, Jansson B, Kierkergaard A, de Wit C (1993) Polybrominated diphenyl ethers (PBDE) in biological samples from the Swedish environment. Chemosphere 26:1703–1718

Shaw SD, Kannan K (2009) Polybrominated diphenyl ethers in marine ecosystems of the American continent: foresight from current knowledge. Rev Environ Health 24:157–229

She J, Petreas M, Winkler J, Visita P, Mckinney M, Kopec D (2002) PBDEs in the San Francisco area: measurements in the harbor seal blubber and human breast adipose tissues. Chemosphere 46:697–707

Skarman E, Darnerud PO, Ohrvik H, Oskarsson A (2005) Reduced thyroxine levels in mice perinatally exposed to polybrominated diphenyl ethers. Environ Toxicol Pharmacol 19:273–281

Slotkin TA, Card J, Infante A, Seidler FJ (2013) BDE99 (2,2′,4,4′,5-pentabromodiphenyl ether) suppresses differentiation into neurotransmitter phenotypes in PC12 cells. Neurotoxicol Teratol 37:13–17

Stapleton HM, Dodder NG, Offenberg JH, Schantz MM, Wise SA (2005) Polybrominated diphenyl ether in house dust and clothes dryer lint. Environ Sci Technol 39:925–931

Stapleton HM, Eagle S, Anthopolos R, Wolkin A, Miranda ML (2011) Associations between polybrominated diphenyl ether (PBDE) flame retardants, phenolic metabolites, and thyroid hormones during pregnancy. Environ Health Perspect 119:1454–1459

Szabo DT, Richardson VM, Ross DG, Diliberto JJ, Kodavanti PR, Birnbaum LS (2009) Effects of perinatal PBDE exposure on hepatic phase I, phase II, phase III, and deiodinase 1 gene expression involved in thyroid hormone metabolism in male rat pups. Toxicol Sci 107:27–39

Ta TA, Koenig CM, Golub MS, Pessah IN, Qi L, Aronov PA (2011) Bioaacumulation and behavioral effects of 2,2′,4,4′-tetrabromodiphenyl ether (BDE-47) in perinatally exposed mice. Neurotoxicol Teratol 33:393–404

Takeshita A, Yen PM, Ikeda M, Cardona GR, Liu Y, Koibuchi N et al (1998) Thyroid hormone response elements differentially modulate the interactions of thyroid hormone receptor with two receptor binding domain in the steroid receptor coactivator-1. J Biol Chem 273:21554–21562

Takeshita A, Taguchi M, Koibuchi N, Ozawa Y (2002) Putative role of the orphan nuclear receptor SXR (steroid and xenobiotic receptor) in the mechanism of CYP3A4 inhibition by xenobiotics. J Biol Chem 277:32453–32458

Tseng L, Li M, Tsai S, Lee C, Pan M, Yao W et al (2008) Developmental exposure to decabromodiphenyl ether (PBDE 209): effects on thyroid hormone and hepatic enzyme activity in male mouse offspring. Chemosphere 70:640–647

Verner MA, Bouchard M, Fritsche E, Charbonneau M, Haddad S (2011) In vitro neurotoxicological data in human risk assessment of polybrominated biphenyl ethers (PBDEs): overview and perspectives. Toxicol In Vitro 25:1509–1515

Viberg H, Fredriksson A, Eriksson P (2002) Neonatal exposure to the brominated flame retardant 2,2′,4,4′,5-pentabromodiphenyl ether causes altered susceptibility in the cholinergic transmitter system in the adult mouse. Toxicol Sci 67:104–107

Viberg H, Fredriksson A, Eriksson P (2003) Neonatal exposure to polybrominated diphenyl ether (PBDE153) disrupts spontaneous behaviour, impairs learning and memory and decreases hippocampal cholinergic receptors in adult mice. Toxicol Appl Pharmacol 192:95–106

Viberg H, Johansson N, Fredriksson A, Eriksson J, Marsh G, Eriksson P (2006) Neonatal exposure to higher brominated biphenyl ethers, hepta-, octa-, or nonabromodiphenyl ether, impairs spontaneous behaviour and learning and memory functions of adult mice. Toxicol Sci 92:211–218

Viberg H, Mundy W, Eriksson P (2008) Neonatal exposure to decabrominated biphenyl ether (PBDE 209) results in changes in BDNF, CaMKII and GAP-43, biochemical substrates of neuronal survival, growth, and synaptogenesis. Neurotoxicology 29:152–159

Williams GR (2008) Neurodevelopmental and neurophysiological actions of thyroid hormone. J Neuroendocrinol 20:784–794

Woods R, Vellaro RO, Golub MS, Suarez JK, Ta TA, Yasui DH, Chi LH et al (2012) Long-lived epigenetic interaction between perinatal PBDE exposure and Mecp308 mutation. Hum Mol Genet 21:2399–2411

Yen PM (2001) Physiological and molecular basis of thyroid hormone action. Physiol Rev 81:1097–1142

Zhang C, Liu X, Chen D (2010) Role of brominated biphenyl ether-209 in the differentiation of neural stem cells *in vitro*. Int J Dev Neurosci 28:497–502

Zhou T, Ross DG, Devito MJ, Crofton KM (2001) Effects of short-term in vivo exposure to polybrominated diphenyl ethers on thyroid hormone and hepatic enzyme activities in weanling rats. Toxicol Sci 61:76–82

Zhou T, Taylor MM, Devito MJ, Crofton KM (2002) Developmental exposure to brominated diphenyl ethers results in thyroid hormone disruption. Toxicol Sci 66:105–116

Chapter 4
Perinatal Infection-Associated Changes in Thyroid Hormone Status, Gut Microbiome, and Thyroid Hormone-Mediated Neurodevelopment

E.M. Sajdel-Sulkowska, M. Bialy, and R. Zabielski

Abstract Maternal/neonatal infections are well recognized and leading causes of infant morbidity and mortality. However, the effect of these infections on thyroid hormone (TH) status and TH-dependent neurodevelopment has been so far overlooked. As adequate TH levels are essential for metabolism, growth, and differentiation and critical for brain development, their deficiency during the perinatal period is likely to result in neurodevelopmental abnormalities. Clinical studies indeed seem to suggest a link between perinatal infection and neurological and neuropsychiatric disorders, but the involvement of TH in their pathology has been relatively understudied. This review addresses the possible link between perinatal infection and abnormal TH-dependent neurodevelopment by posing several basic questions: (1) How frequently are perinatal infections associated with changes in TH status? (2) What are the possible mechanisms involved in the dysregulation of TH status in response to infection? (3) How strong is the evidence for the association of infection-linked neurodevelopmental abnormalities and altered TH status? (4) Are there sex-specific differences in short-term morbidity and neurodevelopmental outcome of neonatal infection? (5) Do TH-dependent therapies present viable prevention/treatment of neurodevelopmental abnormalities associated with perinatal infection?

E.M. Sajdel-Sulkowska, D.Sc. (✉)
Department of Psychiatry, Harvard Medical School and Brigham and Women's Hospital, 75 Francis St., Boston, MA 02114, USA

Department of Experimental and Clinical Physiology, Laboratory of Center for Preclinical Research, Medical University of Warsaw, Warsaw, Poland

Department of Physiological Sciences, Warsaw University of Life Sciences, Warsaw, Poland
e-mail: esulkowska@rics.bwh.harvard.edu

M. Bialy
Department of Experimental and Clinical Physiology, Laboratory of Center for Preclinical Research, Medical University of Warsaw, Warsaw, Poland

R. Zabielski
Department of Physiological Sciences, Warsaw University of Life Sciences, Warsaw, Poland

© Springer Science+Business Media New York 2016
N. Koibuchi, P.M. Yen (eds.), *Thyroid Hormone Disruption and Neurodevelopment*, Contemporary Clinical Neuroscience, DOI 10.1007/978-1-4939-3737-0_4

These questions will be considered while reviewing data of human clinical studies, animal models, and observations of farm animals.

Keywords Perinatal infection • Thyroid hormone • Neurodevelopment • Neuropsychiatric disorder • Sexual dimorphism • Breastfeeding

4.1   Introduction

Maternal/neonatal infections are well recognized and leading causes of infant morbidity and mortality. Worldwide these infections, which include neonatal sepsis/pneumonia, tetanus, and diarrhea, are responsible for 35 % of neonatal deaths (Lozano et al. 2011). The prevalence of neonatal/infant infections is higher in the developing countries when compared to the developed world; in the developing countries, the incidence of sepsis is 10–30 % and mortality at 30–45 % (Vain et al. 2012). While the immediate effects of infections are now identified as a major challenge of the Millennium Development Goal #4 (Lozano et al. 2011), the effects of perinatal infections on the thyroid hormone (TH) status and downstream TH-dependent neurodevelopment have been so far overlooked. THs are essential for metabolism, growth, and differentiation of number of tissues including the stomach and gut. They are critical for brain development, and their deficiency during the perinatal period is likely to result in neurodevelopmental abnormalities. Clinical studies indeed seem to suggest a link between the perinatal infection and the pathology of neurological and neuropsychiatric disorders, but the possible involvement of THs has been relatively understudied.

This review addresses the possible link between perinatal infection and abnormal TH-dependent neurodevelopment by posing several basic questions: (1) How frequently are perinatal infections associated with changes in TH status? (2) What are the possible mechanisms involved in the dysregulation of TH status in response to infection? (3) How strong is the evidence for the association of infection-linked neurodevelopmental abnormalities and altered TH status? (4) Are there sex-specific differences in short-term morbidity and neurodevelopmental outcome of neonatal infection? (5) Do TH-dependent therapies present viable prevention/treatment of neurodevelopmental abnormalities associated with perinatal infection? Furthermore, it considers data suggesting a critical role of the gastrointestinal tract (GIT) and its microbiome in defining host response to infection.

In the context of dysregulation of TH status during perinatal infection, one must recognize an important role of the breastfeeding. Both the colostrum and the milk provide not only TH and iodine (Andersen et al. 2014; de Oliveira et al. 2011; Flachowsky et al. 2014; discussed below) but also several biologically active milk-borne peptides and proteins important for maintaining gut health and controlling microbial ecosystem (Raikos and Dassios 2014; López-Expósito and Recio 2008; Baldi et al. 2005; Makowska et al., 2016). Infection management practices both in humans and animals frequently employ interruption in nursing during infection in neonates. Such practices may further contribute to infant's TH deficiency during infection and have dire consequences on both immediate and long-term health (Makowska et al., 2016).

1. How frequently are perinatal infections associated with changes in TH status?

Thyroid functions are critical to the transition from fetal to neonatal life and important for maturation of many body functions such as thermoregulation and breathing and are critical for neurodevelopment. Several studies examined the state of hypothalamic-pituitary-thyroid axis (HTP) axis in infected newborns, which revealed a relationship between the clinically manifested intrauterine infection and the change in the concentration of TH. A decrease in triiodothyronine (T3) and thyroxine (T4) and an increase in thyrotrophic hormone (TSH) have been observed in umbilical blood of newborns and had a significant correlation with the development of sepsis in newborns (Nikoleishvili et al. 2005). One of the first-of-the-kind studies, involving 292 newborns, showed serum T3 levels in septic newborns were significantly decreased with respect to those of healthy newborns (Kurt et al. 2011).

Among neonatal infections, the neonatal sepsis is characterized by symptoms of infection but may be not accompanied by bacteremia (Fabris et al. 1995). In the study involving 49 neonates 1–4 weeks of age, the neonates with sepsis had a lower mean serum total T4 and T3 as compared to healthy neonates, with non-survivors having a significantly lower T3 values (Das et al. 2002). It was concluded that serum TH levels respond transiently to sepsis and are predictors of adverse outcome in neonates with sepsis. In another study involving 49 neonates, both T3 and T4 levels were significantly decreased in neonates with sepsis as compared to controls, with non-survivors having lower hormonal levels (Sharma et al. 2013). Several studies in critically ill children showed an association of decreased levels of T3 and T4 with mortality (Uzel and Neyzi 1986; Yildizdas et al. 2004).

While the above representative studies suggest an association between HPT axis dysfunction and morbidity and mortality of infection in humans, investigations of HPT axis in critically ill animals are limited. Results of recent studies in critically ill newborn foals indicate that concentration of total and free T4 and T3 was significantly decreased in septic and sick non-septic foals. Reductions in hormone concentration were associated with an increased sepsis score, with non-surviving septic foals having lower T4 and T3 concentrations than surviving septic ones (Himler et al. 2012). A significant decrease in the values of T4 and free T3 was also observed in malignant ovine theileriosis, a fatal disease of sheep caused by the pathogenic species of intracellular protozoans of the genus *Theileria* (Nazifi et al. 2012). Also, it has been reported that TH decreased significantly following experimental (Garg et al. 2001; Sangwan et al. 2003) and natural (Khalil et al. 2011) infections with *T. annulata* in cattle. It is possible that inflammation of the gut mucosa inhibits the absorption of THs and iodine from the gut lumen thereby accelerating the vicious cycle.

As mentioned above, colostrum and milk breastfeeding provides considerable amounts of both TH and iodine in humans (Andersen et al. 2014; de Oliveira et al.

2011) and in cattle (Flachowsky et al. 2014), which supplements infant's developing TH functions. Both THs and iodine in the milk are readily absorbed from the gut lumen into the circulation (Koldovsky et al. 1995), in particular during the first few postnatal days when the gut barrier is open (Tenore et al. 1980). In dairy cattle, milk iodine content is determined by the level of iodine supplementation in the feed, iodine source, the presence of iodine antagonists such as glucosinolates in the feed, farm management, the nature of teat dipping with iodine-containing substances, and milk processing in the dairy (Flachowsky et al. 2014). Colostral and breast milk THs in synchrony with other milk-borne hormones (leptin, ghrelin, opiates, gut regulatory peptides) and growth factors (EGF, IGFs) are responsible for postnatal maturation of the GIT mucosa and thereby are indirectly responsible for gut microbial ecosystem development (Kapica et al. 2008; Zabielski et al. 2008). They may presumably be also responsible for local nervous system development (Zabielski 2007). Interestingly, rat studies have shown that maternal hyperleptinemia induced by high-energy diet may stimulate both the adrenal medullary and the thyroid function during the lactation period (Franco et al. 2012), which may result in accelerated maturation of the GIT in the neonatal offspring, but abnormal TH function in weaned pups (Passos et al. 2012) and cardiovascular diseases and other signs of metabolic syndrome in adulthood (Franco et al. 2012). However, studies in neonatal calves showed that extended colostrum feeding from birth to 1 month of age was unsuccessful in modifying the endogenous profile of THs in blood plasma (Babitha et al. 2011).

Importantly, the composition of milk with respect to both TH and iodine can be affected by various environmental factors (Andersen et al. 2014; de Oliveira et al. 2011; Flachowsky et al. 2014). For example, breast milk TH levels are decreased by nicotine exposure (de Oliveira et al. 2011), while high dietary iodine intake of lactating women in Korean population resulted in increased iodine level in breast milk (Chung et al. 2009), but the increased iodine level correlated with a frequency of subclinical hypothyroidism in preterm infants. These data suggest that caution must be applied when considering iodine supplementation programs in preterm infants (Wang et al. 2009). On the other hand, TH supplementation seems to be effective in reducing gastrointestinal disorders, improving body weight gains and increasing quantity of tolerated milk in intrauterine growth-retarded neonates (Komiyama et al. 2009).

From the above discussion, it can be concluded that in addition to the endogenous factors, breastfeeding plays an important role in regulating TH status in infected neonates, as illustrated in Fig. 4.1. However, present pediatric and veterinary infection management practices frequently include interruption in nursing during infection in neonates and replacement of mother's milk with formulas containing iodine as a supplement but lacking THs. In addition, soybean proteins replacing human/animal proteins in the milk interfere with TH absorption (Fruzza et al. 2012). Such practices may further contribute to infant's TH deficiency and impact neurodevelopment.

Moreover, both colostrum and milk contain important quantities of biologically active milk-borne peptides and proteins known to stimulate the offspring's immune

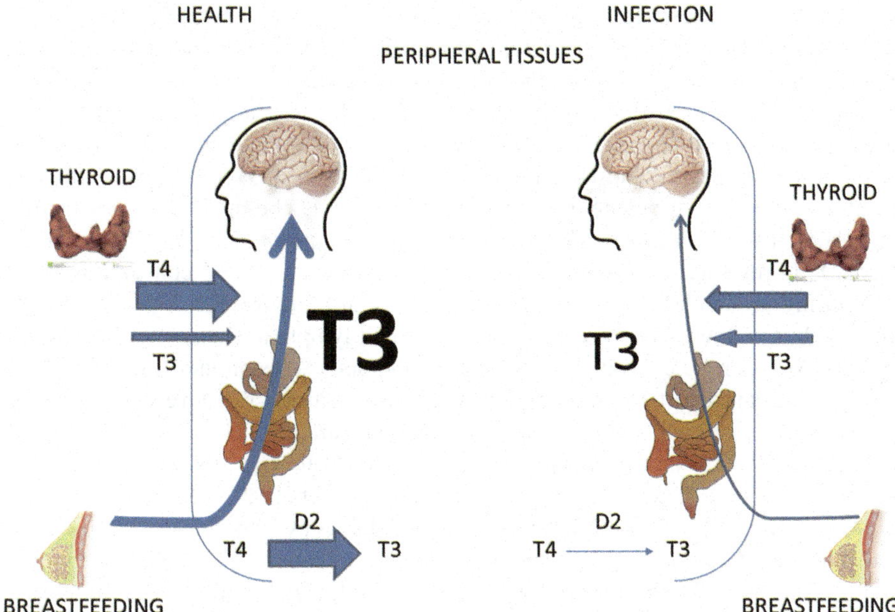

Fig. 4.1 Endogenous and exogenous factors affecting thyroid hormone status in healthy vs. infected neonates

system, digestion, and absorption of nutritional elements; development of endogenous defense mechanisms against bacteria, fungi, and viruses; prebiotic effects; and others (Raikos and Dassios 2014; López-Expósito and Recio 2008; Baldi et al. 2005). Thus eliminating nursing during infection may exacerbate clinical symptoms of infection and have dire consequences on both immediate and downstream health.

2. **What are the possible mechanisms involved in the dysregulation of TH in response to infection?**

Lower levels of T3 and T4 could partly be due to the anorexia condition, prevailing during the infection and associated with both reduced milk feeding and reduced absorption of THs, but can hardly explain the overall effect. TH levels could also be affected by altered levels of various lymphokines which influence HPT axis modulating either the thyroid hormone levels or cytokine production by thyrocytes (Fabris et al. 1995). Decreased TH levels could affect the integrity of gut barrier against further pathogenic influx (Yang et al. 2003), reduce stomach acid levels and facilitate dysbiosis (an imbalance between pathogenic and beneficial bacteria in the gut), and impair the clearance of hormones like estrogen which can impact thyroid hormone action further. Decreased TH levels in neonatal rats led to alterations in carbohydrate metabolism, which increased susceptibility to the development of glucose intolerance and occurrence of type 2 diabetes later in life (Farahani et al. 2013).

Relatively unrecognized is a function of the gastrointestinal tract (GIT) in the regulation of TH status. Healthy gut converts T4 to T3 locally within its tissues as suggested by the presence of the type 2 iodothyronine deiodinase (D2; Bates et al. 1999). While T4 is the main secretory product of the thyroid gland, T3—the main metabolically active TH—is produced enzymatically from T4 locally in many tissues including the intestine; the local production of T3 in the intestine represents, however, no more than circa 6 % of the whole body T3 (Nguyen et al. 2004). The deiodinase enzymes (including D1, D2, and, D3) are essential cellular determinants of intracellular activation and deactivation of TH with D2 being directly involved in converting inactive thyroxine (T4) to the active triiodothyronine (T3). Altered activity of D2 in response to inflammation would thus affect T4 to T3 conversion in the GIT during infection (Barca-Mayo et al. 2011). Furthermore, the digestive tract is lined with immune tissue known as gut-associated lymphoid tissue (GALT). Stress to the GALT can be caused by a number of factors (Acheson and Luccioli 2004) with the increased cortisol impacting TH functions, raising rT3 level and lowering T3 levels. Also cell walls of pathogenic bacteria can affect TH by reducing levels of T3, increasing rT3, and decreasing TSH (van der Poll et al. 1999) and reducing thyroid hormone receptor (TR) expression (Beigneux et al. 2003).

Interestingly, the gut bacteria are also involved in converting inactive T4 into the active hormone T3 and may be the key contributor to neonatal TH levels during the first few days after birth in a healthy neonate while the gut epithelium is immature. Around 20 % of T4 in the body is converted to T3 in the GIT, in the forms of T3 sulfate (T3S) and triidothyroacetic acid (T3AC). The conversion of T3S and T3AC into active T3 requires an enzyme called intestinal sulfatase produced by healthy gut bacteria. Intestinal dysbiosis significantly inhibits the conversion of T3S and T3AC to T3. Inflammation that arises out of dysbiosis creates more demand for cortisol which is required for both the maturation of the gut epithelium and the reduction on the inflammation including infections from bacteria, yeast, and parasites, which in turn raises the cortisol production by the adrenal gland. Cortisol causes a shift in TH metabolism increasing T4 and causing imbalance. Chronic elevation of cortisol suppresses the immune system leading to dysbiosis and creating a vicious cycle and further disruption of TH. Interestingly, recent studies have shown that intestinal inflammation (necrotizing enterocolitis) can be counteracted by milk that contains many proteins with anti-inflammatory properties (Chatterton et al. 2013). This evidence further argues for continuing the nursing practice during neonatal infection.

The maturation of the GIT in preterm animals, including gut permeability, is prevented by the toxic effect of external factors including bacterial translocation. Many toxins, toxicants, diet, stress, and infectious agents affect an individual's gut microbiome, which may be especially sensitive during the critical developmental period. Disruption of the developing microbiome may have profound consequences on the developing gut–brain axis including the brain as well as long-term consequences on the neurodevelopment (Sajdel-Sulkowska and Zabielski 2013). As has been pointed out (Sajdel-Sulkowska and Zabielski 2013), the microbiome responsible for the local activation of TH could communicate changes in the TH status to the brain by a direct neuronal pathway via the vagus nerve, the hormonal pathway of several hormones involved in the regulation of metabolism including the TH, and the immunological signaling pathway involving the cytokines. Recent studies suggest that the vagus nerve is involved in immunomodulation as suggested by its ability to attenuate the production of pro-inflammatory cytokines in experimental models of inflammation (de Jonge and Ullola 2007). Furthermore, the gut microbiome emerges as a major player not only in the maturation of the GIT tissue and the gut–brain axis but also in brain maturation (Sajdel-Sulkowska et al. 2015), through its effect on both the immune and endocrine systems. A recent clinical study suggests a cross talk between the HPT axis and the gut and a significant association of gut hormones with TH (Emami et al. 2014).

3. How strong is the evidence for the association of infection-linked neurodevelopmental abnormalities and altered TH status?

TH plays a critical role during neurodevelopment. In humans, TH is essential for optimal brain development during gestation and for the first 2 years postnatally. Maintaining adequate TH levels is vital, as low levels (even transiently low) of the hormone are associated with adverse neurodevelopmental outcome. Importantly, the developmentally regulated TH levels are decreased by infection (Williams et al. 2013). Decreased TH levels may contribute to increased morbidity and mortality of infections during the perinatal period and especially in the premature infants and affect their neurodevelopment. For example, infections with candida in premature infants have a much higher rate of mortality and neurodevelopmental disabilities (60 % vs. 28 %) than term controls (Lee et al. 1998). Infants who acquire HIV during fetal and early neonatal life tend to display poorer mean developmental scores than HIV-unexposed children with mean motor and cognitive scores consistently 1–2 SDs below the population mean (Le Doare et al. 2012). Data suggest that maternal influenza infection is associated with a twofold increased risk of infantile autism (Atladottir et al. 2012).

While both clinical and epidemiological data suggest that infection during pregnancy and nursing increases the probability of neonatal brain injury and may have a long-lasting impact on brain functions, it fails to recognize the potentially critical impact of altered TH status during infection on neurodevelopment.

Maternal infection during pregnancy affects the health of both the mother and the unborn child and has been linked to neurological and neuropsychiatric disorders in humans such as cerebral palsy (Schendel et al. 2001; Schendel 2002), neonatal strokes (Ferriero 2004), Parkinson's disease (Carvey et al. 2003) schizophrenia (Watson et al. 1999; Pearce 2001), and affective disorders (Watson et al. 1999). Animal studies implicate bacterial infection in the pathology of Parkinson's disease (Carvey et al. 2003) and autism (Patterson 2002). The link between neonatal infection and neurological and behavioral disorders is presented in Fig. 4.2.

It has been hypothesized that maternal infection affects the developing brain, and several studies have confirmed the developmental impact of lipopolysaccharides (LPSs), major components of the outer membrane of Gram-negative bacteria, on the brain and behavior (Ghiani et al. 2011). Additionally, LPS challenge has been reported to be associated with decreased neurogenesis (Cui et al. 2009), increased apoptosis (Sharangpani et al. 2008), and disruption of the blood–brain barrier (Stolp et al. 2005).

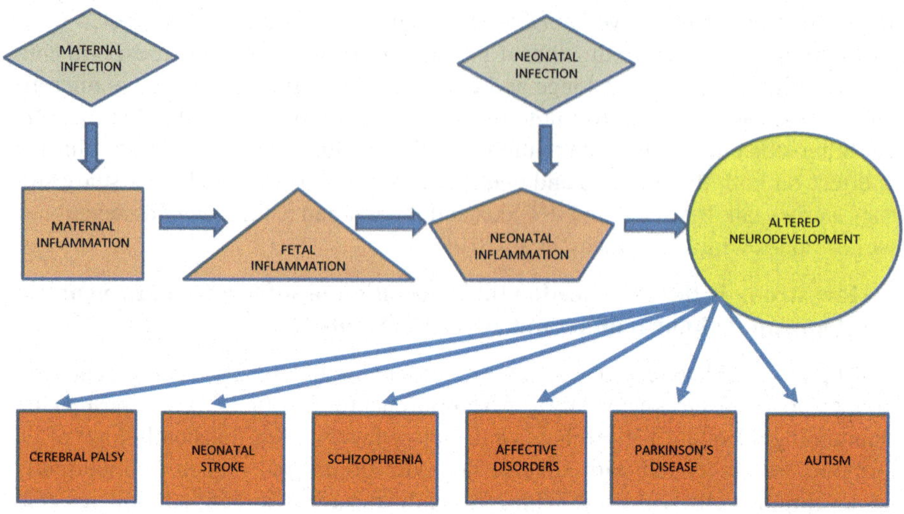

Fig. 4.2 The link between neonatal infection and neurological and behavioral disorders

LPS model bacterial infection in animals by triggering a response that involves the production of inflammatory mediators—cytokines. LPS administered to the pregnant mother are transferred to the fetus through the placenta (Kohmura et al. 2000) and result in increased cytokines levels in amniotic fluid (Urakubo et al. 2001; Gayle et al. 2004) and the fetal brain (Urakubo et al. 2001). Bacterial infection of the lactating mother also results in increased levels of cytokines in the milk (Bannerman et al. 2004). LPS accumulation in the fetal brain (Urakubo et al. 2001) suggests that it can directly affect both fetal and neonatal CNS development and behavior.

LPS also triggers increased oxidative stress (Cambonie et al. 2004; Gayle et al. 2002; Rohl et al. 2010). Experimental evidence suggests that cellular injury during the inflammatory response is mediated by reactive oxygen species (Davies 2000) and that endotoxin-induced excessive free radical production damages and opens the blood–brain barrier (Gaillard et al. 2003). Further, LPS is not only a potent trigger of oxidative stress (Cambonie et al. 2004; Gaillard et al. 2003) but also a disruptor of antioxidant defenses (Cambonie et al. 2004). The combined effect of high levels of free radicals and low levels of scavengers leads to oxidative stress (Sebai et al. 2009) that promotes injury in the developing brain.

LPS also regulates expression of D2, an enzyme responsible for most of the T3 supply within the brain (Silva 1983). LPS-mediated increase in D2 in glial cultures has been shown to be associated with elevated T3-dependent gene expression in cocultured neurons (Freitas et al. 2010), suggesting that LPS-induced changes in D2 or oxidative stress could impact the developing CNS. Interestingly, LPS exerts a sex-dependent response (Engeland et al. 2003), and LPS-induced cytokine expression exhibits a sexually dimorphic profile (Nguyen et al. 2009; Gourdy et al. 2005).

In the recently completed study, we examined the effect of perinatal exposure to *E. coli* LPS during critical developmental periods on the developing rat brain employing animal model of infection (Xu et al. 2013). LPS exposure is one of the most acceptable models of infection and a sufficient trigger for cytokine production. Our study was undertaken to address the hypothesis that maternal infection during pregnancy leads to increased oxidative stress which in turn inhibits cerebellar D2 activity and results in an altered pattern of TH-dependent gene expression and impaired cerebellar development. To test this supposition, we examined the effects of maternal challenge with LPS in two strains of rats with different thresholds to oxidative stress, spontaneously hypertensive rats (SHRs) and Sprague–Dawley (SD) rats, to test for genetically dependent sensitivity to inflammation, and in the male and in female neonates, to test for the sex-dependent nature of these effects. Neurodevelopment was assessed in terms of milestones such as rollover time, auditory function, and motor learning. The neurodevelopmental findings were correlated with cerebellar levels of the oxidative stress marker 3-nitrotyrosine (3-NT) and D2 activity. We further examined the expression of several TH-dependent genes and genes implicated in cerebellar development. Our data suggest that maternal infection exerts a negative neurodevelopmental impact in the offspring and that the effect appears to be both strain and sex dependent. Thus, maternal exposure to LPS during the perinatal period leads to deficiencies in motor learning, which are manifested in a strain- and sex-dependent manner. These changes are accompanied by a selective increase in oxidative stress, decreased D2 activity, and altered expression of cerebellar genes (Xu et al. 2013; Fig. 4.2). Alterations in gene expression could be, in part, the result of local changes in TH availability. While the mechanism by which D2 activity is decreased in LPS-challenged pups is unclear, it is possible that it could be modulated by low selenium levels accompanying sepsis both in terms of a direct effect on the production of the D2, a selenoenzyme, and also in terms of susceptibility of the cell to oxidative stress (Forceville 2007).

Other studies have shown that perinatal LPS exposure has a profound effect on the GIT function as LPS enhances GIT motility (speeds up digesta transit), reduces absorption of bioactive substances from the food, and exerts an effect on microbiome similar to the repeated treatment with antibiotics. Developmentally abnormal gut microbiome may in turn affect both the gut–brain axis and brain development and contribute to the etiology of a number of neuropsychiatric disorders (rev. Sajdel-Sulkowska et al. 2015). Experiments in healthy mice have shown that disrupting balance of gut microbiome with antibiotics caused changes in mice behavior and was accompanied by changes in BDNF which has been linked to depression and anxiety (Bercik et al. 2011; Neufeld et al. 2011). Perinatal LPS exposure affects gut motility as suggested by studies of irritable bowel syndrome (IBS), where mild bacterial overgrowth-associated motility disorder can be reversed by antimicrobials (Scarpignato and Pelosini 1999). Animal studies have also shown that stress can change the composition of microbiome, where the changes are associated with increased vulnerability to inflammatory stimuli in the GIT. Could gut dysbiosis be induced by recurrent infections? Indeed, inflammation of the gut negatively affects the butyrogenic bacteria and butyrate available for the microbial ecosystem, the gut, and the entire organism, further affecting health status (Guilloteau et al. 2010).

We have observed an increase in neurotrophin levels in the cerebella of rats exposed to LPS (Sajdel-Sulkowska, unpublished observation) and brain region-specific changes in neurotrophin levels in autism spectrum disorder (ASD; Sajdel-Sulkowska and Koibuchi 2011; rev. Sajdel-Sulkowska et al. 2015). Together these observations suggest that a bacterial infection could trigger gut microbiome to induce cytokine overproduction leading to imbalance of brain neurotrophins and contribute to developmental abnormalities.

Gut microbiome, regulated by both intrinsic and extrinsic factors, may be further jeopardized by recurrent infections and/or recurrent use of antibiotics. Developmentally abnormal gut microbiome may in turn affect both gut–brain axis and brain development and contribute to the etiology of a number of neuropsychiatric disorders including autism (Sajdel-Sulkowska et al. 2015).

Animal studies have shown that stress can change the composition of microbiome, where the changes are associated with increased vulnerability to inflammatory stimuli in the GIT (Gareau et al. 2006); microbiome plays here an important role in memory dysfunction (Gareau et al. 2011). Stress is known to inhibit gut contraction, one of the crucial defense strategies against bacterial colonization of the gut mucosa.

4. Are there sex-specific differences in short-term morbidity and neurodevelopmental outcome of neonatal infection?

Sexual dimorphism of CNS structure, function, and response to environmental perturbations has been observed in humans and animals (Nguon et al. 2004, 2005; Sulkowski et al. 2012; Khan et al. 2012; Xu et al. 2013) but remains relatively unrecognized in the context of neonatal infection. Importantly, most of the neurodevelopmental disorders afflict a significantly larger proportion of males than females (Rutter et al. 2003). Furthermore, the neuropsychiatric disorders linked to neonatal infections, such as autism and Parkinson's disease, also afflict a significantly larger proportion of males, with a male-to-female ratio 4:1 for autism (Baird et al. 2001; Scott et al. 2002) and 2:1 ratio for Parkinson's disease (Van Den Eaden et al. 2003).

Nevertheless, evidence derived from both animal studies and clinical observations supports the notion that sex influences host immune function (Klein et al. 2002) with male having a higher antibody response to a viral infection (Klein et al. 2002). Other studies support sex-specific differences in innate antiviral responses that may be determined by gonadal hormones (Hannah et al. 2008). Furthermore, neonatal infection elicits sex-specific growth response and may result in sex-specific neurodevelopmental outcome. In animals, LPS exerts a sex-dependent response (Engeland et al. 2003). Furthermore, LPS-induced cytokine expression exhibits a sexually dimorphic profile (Nguyen et al. 2009; Gourdy et al. 2005). In a mouse model, intrauterine inflammation has been shown to be associated with acute brain insult and changes in MRI and behavior throughout the neonatal period and adulthood. Furthermore, brain inflammation in the offspring involves microglial activation, increase in the number of microphages in the brain, and neuronal loss, in a sex-specific manner (Dada et al. 2014). Our own animal study showed that maternal exposure to LPS during the perinatal period leads to deficiencies in motor learning in the offspring, which are manifested in a strain- and sex-dependent manner. These changes are accompanied by a selective increase in oxidative stress, decreased D2 activity, and altered expression of cerebellar genes (Xu et al. 2013; Fig. 4.2). In another study, neonatal infection in males induces upregulation of genes associated with cellular and nervous system development and function (Wynne et al. 2011).

Interestingly, congenital cytomegalovirus infection in humans is associated with intrauterine growth restriction (IUGR) due to impaired placental development and functions (Pereira et al. 2014). IUGR neonates in turn have a higher mortality and morbidity as compared to neonates born with normal weight (Aucott et al. 2004; Che et al. 2010). Placental disease predisposes the severely growth-restricted neonate to inflammation and subsequently to necrotizing enterocolitis (Aucott et al. 2004; Patole 2007; Manogura et al. 2008), although these findings could not be confirmed in IUGR pig model (Che et al. 2010). Some have suggested TH dysregulation in IUGR (Mahajan et al. 2005). Several studies also suggested sexually dimorphic growth in IUGR animals (Oyhenart et al. 2003); both somatic and cerebral growth in IUGRs showed a sex-specific pattern that resulted in abolition of the sexually dimorphic growth curve observed in control pups (Dressino et al. 2002). Sex-related growth differences were also observed in IUGR piglets that included abnormalities in the kidneys (Wlodek et al. 2007), gut, pancreas, and brain (Mickiewicz et al. 2012). In another study, a reduction in nephron number that led to hypertension was only observed in IUGR male but not female rats (Wlodek et al. 2007). Sexual dimorphism was also observed in skeletal growth which was faster in females than in male IUGR rats; however, the level of bone mineralization was higher in females than male IUGR rats (Romano et al. 2009). Furthermore, the pancreas of IUGR piglets which was in general less flexible in adaptation to dietary changes as compared to non-IUGR littermates was further affected in female IUGRs as indicated by lower organ weight, total protein, and lipase and amylase activities as compared to male IUGRs and did not respond to dietary stimulation. Furthermore, small intestinal relative length and mucosal thickness in IUGR females were smaller as compared to IUGR males (Mickiewicz et al. 2012).

Transitioning from findings in animal models, clinical studies involving 45 women indicated that male fetuses mounted a larger pro-inflammatory response to LPS (Kim-Fine et al. 2012). Girls less frequently suffered from early-onset sepsis, but boys did not suffer greater adverse neurodevelopmental problems at 12 or 24 months (Neubauer et al. 2012). Furthermore, male sex predisposes to severe sepsis and septic shock with infant boys between 1 and 12 months being almost three times more likely to succumb to sepsis (Bindl et al. 2003).

The above observations suggest that neonatal sex influences not only magnitude of host immune response but also elicits sex-specific growth response and may result in sex-specific neurodevelopmental outcome. Interestingly, breastfeeding may confer protection against respiratory infections in girls but not in boys (Sinha et al. 2003). This observation, if confirmed, would suggest that sex of the neonates may be an important factor in infection management.

5. Do TH-dependent therapies present viable prevention/treatment of neurodevelopmental abnormalities associated with perinatal infection?

Clinical, biochemical, and immunological evidence and animal data suggest that TH would be beneficial in treatment of infection by itself or as an adjuvant to present-day therapies. THs are the only consistent natural stimulator of both the primary lymphoid tissue (thymus and bone marrow) and secondary lymphoid tissue (circulating lymphocytes, spleen, and lymph nodes). There is abundant historical precedent for using thyroid hormone as a safe pharmacological agent against human immunodeficiency virus disease (Derry 1996).

The role of T3 supplement in protecting gut barrier has been studied in septic rat model. Serum free T3 and T4 concentrations were lower than in sham group, but it was corrected in sepsis group supplemented with T3. Furthermore, transmission electron microscopy (TEM) and light microscopy showed that T3 supplements preserved ultrastructure and morphology of intestinal mucosa in septic rats suggesting that T3 protects gut barrier (Yang et al. 2003). Recent study compared the therapeutic effects of methylprednisolone (MP) and T3 replacement therapy during an early sepsis (Coskun et al. 2012) in Wistar rats. While a septic insult resulted in significant alterations in free T3, free T4, and cortisol levels, compared to the MP replacement therapy, therapeutic effects of T3 replacement therapy have been found significantly more promising (Coskun et al. 2012).

While more research is needed to better understand the mechanisms involved in the HPT axis and specifically TH response to infection, manipulation of HPT axis or TH therapy should be seriously considered in humans and in animals, including farm animals.

And finally, reinstatement of uninterrupted nursing during infection should be considered in managing the neonatal infection. Recent studies suggest that intestinal inflammation (necrotizing enterocolitis) can be counteracted by milk that contains many proteins with anti-inflammatory properties (Chatterton et al. 2013). Thus breastfeeding may not only provide both TH and iodine supplementation, but may be beneficial in couneracting inflammation.

4.2   Conclusions

This review addresses the possible relationship between perinatal infections, disrupted TH status, and abnormal TH-dependent neurodevelopment. Both clinical and experimental animal data suggest a link between perinatal infection and neurological and neuropsychiatric disorders, but the possible involvement of TH has been relatively understudied. Evidence presented here points to an association between perinatal infection and decrease in TH levels. The possible mechanism(s) may involve a decrease in local TH activation by the GIT as well as TH activation by microbiome. The neonatal sex may be an important factor in short-term response and long-term neurodevelopmental impact of neonatal infection. While more research is needed to better understand the mechanisms involved in the HPT axis and specifically TH response to infection, hormone replacement therapy should be seriously considered in both pediatric and veterinary management of neonatal infections in humans and in animals, including farm animals.

References

Acheson DW, Luccioli S (2004) Microbial-gut interactions in health and disease. Mucosal immune responses. Best Pract Res Clin Gastroenterol 18:387–404

Andersen SL, Moller M, Laurberg P (2014) Iodine concentrations in milk and in urine during breastfeeding are differently affected by maternal fluid intake. Thyroid 24(4):764–772 (Epub PMID: 24199933)

Atladottir HO, Henriksen TB, Schendel DE (2012) Autism after infection, febrile episodes, and antibiotic use during pregnancy: an exploratory study. Pediatrics 130:1447–1454

Aucott SW, Donohue PK, Northington FJ (2004) Increased morbidity in severe early intrauterine growth restriction. J Perinatol 24:435–440

Babitha V, Philomina PT, Dildeep V (2011) Effect of extended colostrum feeding on the plasma profile of insulin, thyroid hormones and blood glucose of crossbred pre-ruminant calves. Indian J Physiol Pharmacol 55:139–146

Baird G, Charman T, Baron-Cohen S (2001) A screening instrument for autism at 18 months of age; a 6-year follow-up study. J Am Acad Child Adolesc Psychiatry 40:737–738

Baldi A, Ioannis P, Chiara P, Eleonora F et al (2005) Biological effects of milk proteins and their peptides with emphasis on those related to the gastrointestinal ecosystem. J Dairy Res 72:66–72

Bannerman DD, Paape MJ, Lee JW et al (2004) Escherichia coli and Staphylococcus aureus elicit differential innate immune responses following intramammary infection. Clin Diagn Lab Immunol 11:463–472

Barca-Mayo L, Liao XH, DiCosmo C et al (2011) Role of type 2 deiodinase in response to acute lung injury (ALI) mice. Proc Natl Acad Sci U S A 108:E1321–E1329

Bates JM, St. Germain DL, Galton VA (1999) Expression profiles of three iodothyronine deiodinases, D1, D2 and D3, in the developing brain. Endocrinology 140:844–851

Beigneux AP, Moser AH, Shigenaga JK et al (2003) Sick euthyroid syndrome is associated with decreased TR expression and DNA binding in mouse liver. Am J Physiol Endocrinol Metab 284:E228–E236

Bercik P, Denov E, Collins J (2011) The intestinal microbiota affect central levels of brain—derived neurotrophic factor and behavior in mice. Gastroenterology 141:599–609

Bindl L, Buderus S, Dablem P et al (2003) Gender-based differences in children with sepsis and ARDS; the ESPNIC ARDS Database group. Intensive Care Med 29:1770–1773

Cambonie G, Hirbee H, Michaud M et al (2004) Prenatal infection obliterates glutamate-related protection against hydroxyl radicals in neonatal brain. J Neurosci Res 75:125–132

Carvey PM, Chang Q, Lipton JW (2003) Prenatal exposure to the bacteritoxin lipopolysaccharide leads to long-term losses of dopamine neurons in offspring: a potential, new model of Parkinson's disease. Front Biosci 8:s826–s837

Chatterton DE, Nguyen DN, Bering SB (2013) Anti-inflammatory mechanisms of bioactive milk proteins in the intestine of newborns. Int J Biochem Cell Biol 45:1730–1747

Che L, Thymann T, Bering SB et al (2010) IUGR does not predispose to necrotizing enterocolitis or compromise postnatal intestinal adaptation in preterm pigs. Pediatr Res 67:54–59

Chung HR, Shin CH, Yang SW et al (2009) Subclinical hypothyroidism in Korean preterm infants associated with high levels of iodine in breast milk. J Clin Endocrinol Metab 94:4444–4447

Coskun F, Saylam B, Kulah B et al (2012) Comparison of the therapeutic effects of tri-iodothyronine and methylprednisolone during early sepsis in laboratory animals. Bratisl Lek Listy 113:339–346

Cui K, Ashdown H, Luheshi GN et al (2009) Effects of prenatal immune activation on hippocampal neurogenesis in the rat. Schizophr Res 113:288–297

Dada T, Rosenzweig JM, Al Shammary M et al (2014) Mouse model of intrauterine inflammation: sex-specific differences in long-term neurologic and immune sequelae. Brain Behav Immun 38:142–150. doi:10.1016/j.bbi.2014.01.014

Das BK, Agarwal JK, Mishra OP (2002) Serum cortisol and thyroid hormone levels in neonates with sepsis. Indian J Pediatr 69:663–665

Davies KJA (2000) An overview of oxidative stress. IUBMB Life 50:241–244

de Jonge WJ, Ullola L (2007) The alpha7 nicotinic acetylcholine receptor as a pharmacological target for inflammation. Br J Pharmacol 151:915–929

de Oliveira E, de Moura EG, Santos-Silva AP (2011) Neonatal hypothyroidism caused by maternal nicotine exposure is reversed by higher T3 transfer by milk after nicotine withdraw. Food Chem Toxicol 49:2068–2073

Derry DM (1996) Thyroid hormone therapy in patients infected with human immunodeficiency virus: a clinical approach to treatment. Med Hypotheses 47:227–233

Dressino V, Orden B, Oyhenart EE (2002) Sexual responses to intrauterine stress: body and brain growth. Clin Exp Obstet Gynecol 29:100–102

Emami A, Nazem R, Hedayati M (2014) Is association between thyroid hormones and gut peptides, ghrelin and obestatin, able to suggest new regulatory relation between the HPT axis and gut? Regul Pept 189:17–21. doi:10.1016/j.regpep.2014.01.001

Engeland CG, Kavaliers M, Ossnkopp KP (2003) Sex differences in the effects of muramyl dipeptide and lipopolysaccharide on locomotor activity and the development of behavioral tolerance in rats. Pharmacol Biochem Behav 74:433–447

Fabris N, Mocchegiani E, Provinciali M (1995) Pituitary-thyroid axis and immune system: a reciprocal neuroendocrine-immune interaction. Horm Res 43:29–38

Farahani H, Ghasemi A, Roghani M, Zahediasl S (2013) Effect of neonatal hypothyroidism on carbohydrate metabolism, insulin secretion, and pancreatic islets morphology of adult male offspring in rats. J Endocrinol Invest 36:44–49

Ferriero DM (2004) Neonatal brain injury. N Engl J Med 351:1985–1995

Flachowsky G, Franke K, Meyer U et al (2014) Influencing factors on iodine content of cow milk. Eur J Nutr 53:351–365

Forceville X (2007) Effects of high doses of selenium, as sodium selenite, in septic shock patients a placebo-controlled, randomized, double blind, multi-center phase II study—selenium and sepsis. J Trace Elem Med Biol 21:62–65

Franco JG, Fernandes TP, Rocha CP et al (2012) Maternal high-fat diet induces obesity and adrenal and thyroid dysfunction in male rat offspring at weaning. J Physiol 590:5503–5518

Freitas BC, Gereben B, Castillo M et al (2010) Paracrine signaling by glial cell-derived triiodothyronine activates neuronal gene expression in the rodent brain and human cells. J Clin Invest 120:2206–2217

Fruzza AG, Demeterco-Berggren C, Jones KL (2012) Unawareness of the effects of soy intake on the management of congenital hypothyroidism. Pediatrics 130:699–722

Gaillard PJ, de Boer AB, Breimer DD (2003) Pharmacological investigations on LPS-induced permeability changes in the blood-brain barrier in vitro. Microw Res 65:24–31

Gareau MG, Jury J, Yang PC et al (2006) Neonatal maternal separation causes colonic dysfunction in rat pups including impaired host resistance. Pediatr Res 59:83–88

Gareau MG, Wine E, Rodrigues DM et al (2011) Bacterial infection causes stress-induced memory dysfunction in mice. Gut 60:307–317

Garg SL, Rose MK, Agraval VK (2001) Plasma cortisol and thyroid hormone concentrations in crossbred calves affected with theileriosis. Indian Vet J 78:583–585

Gayle DA, Ling Z, Tong C (2002) Lipopolysaccharide (LPS)-induced dopamine cell loss in culture: roles of tumor necrosis factor-alpha, interleukin-1beta, and nitric oxide. Brain Res Dev Brain Res 133:27–35

Gayle DA, Beloosesky R, Desai M et al (2004) Maternal LPS induces cytokines in the amniotic fluid and corticotropin releasing hormone in the fetal rat brain. Am J Physiol Regul Integr Comp Physiol 286:R1024–R1029

Ghiani CA, Mattan NS, Nobuta H et al (2011) Early effects of lipopolysaccharide-induced inflammation on foetal brain development in rat. ASN Neuro. 3(4). pii: e00068. doi:10.1042/AN20110027

Gourdy P, Araujo LM, Zhu R (2005) Relevance of sexual dimorphism to regulatory T cells, estradiol promotes IFN-(gamma) production by invariant natural killer T cells. Blood 105:2415–2420

Guilloteau P, Martin L, Eeckhaut V et al (2010) From the gut to the peripheral tissues: the multiple effects of butyrate. Nutr Res Rev 23:366–384

Hannah MF, Bajic VB, Klein SL (2008) Sex differences in the recognition of and innate antiviral responses to Seoul virus in Norway rats. Brain Behav Immun 22:503–516

Himler M, Hurcombe SD, Griffin A et al (2012) Presumptive nonthyroidal illness syndrome in critically ill foals. Equine Vet J Suppl 41:43–47

Kapica M, Puzio I, Kato I, Kuwahara A, Zabielski R (2008) Role of feed-regulating peptides on pancreatic exocrine secretion. J Physiol Pharmacol 59:145–160

Khalil B, Khodadad M, Pourjafar M et al (2011) Serum thyroid hormones and trace element concentrations in crossbred holstein cattle naturally infected with Theileria annulata. Comp Clin Path 20:115–120

Khan A, Sulkowski ZL, Chen T et al (2012) Sex-dependent changes in cerebellar TH levels and TH-dependent gene expression following perinatal exposure to thimerosal. J Physiol Pharmacol 63:277–283

Kim-Fine S, Regnault TR, Lee JS et al (2012) Male gender promotes an increased inflammatory response to lipopolysaccharide in umbilical vein blood. J Matern Fetal Neonatal Med 25:2470–2474

Klein SL, Marson AL, Scott AL et al (2002) Neonatal sex steroids affect responses to Seoul virus infection in male but not female Norway rats. Brain Behav Immun 16:736–746

Kohmura Y, Kirikae T, Kirikae F et al (2000) Lipopolysaccharide (LPS)-induced intra-uterine fetal death (IUFD) in mice is principally due to maternal cause but not fetal sensitivity to LPS. Microbiol Immunol 44:897–904

Koldovsky O, Illnerova H, Macho L et al (1995) Milk-borne hormones: possible tools of communication between mother and suckling. Physiol Res 44:349–351

Komiyama M, Takahashi N, Yada Y et al (2009) Hypothyroxinemia and effectiveness of thyroxin supplementation in very low birth weight infants with abdominal distension and poor weight gain. Early Hum Dev 85:267–270

Kurt A, Aygun AD, Sengul I (2011) Serum thyroid hormone levels are significantly decreased in septic neonates with poor outcome. J Endocrinol Invest 34:92–96

Le Doare K, Bland R, Newell ML (2012) Neurodevelopment in children born to HIV-infected mother by infection and treatment status. Pediatrics 130:1326–1344

Lee BE, Cheung PY, Robinson JL, Evanochko C (1998) Comparative study of mortality and morbidity in premature infants (birth weight, <1,250 g) with candidemia or candidal meningitis. Clin Infect Dis 27:559–565

López-Expósito I, Recio I (2008) Protective effect of milk peptides: antibacterial and antitumor properties. Adv Exp Med Biol 606:271–293

Lozano R, Wang H, Foreman KJ et al (2011) Progress towards Millennium Goals 4 and 5 on maternal and child mortality: an updated systematic analysis. Lancet 378:1139–1165

Mahajan SD, Aalinkal R, Singh S (2005) Thyroid hormone dysregulation in intrauterine growth retardation associated with maternal malnutrition and/or anemia. Horm Metab Res 37:633–640

Makowska M, Kasarello K, Bialy M, Sajdel-Sulkowska EM, (2016) Autism: "Leaky gut", prematurity and lactoferrin. Ausin J Autism & Relat Disabil. 2(3):idl1021

Manogura AC, Turan O, Kush ML (2008) Predictors of necrotizing enterocolitis in preterm growth-restricted neonates. Am J Obstet Gynecol 198:638.e1–e5

Mickiewicz M, Zabielski R, Grenier B et al (2012) Structural and functional development of small intestine in intrauterine growth retarded porcine offspring born to gilts fed diets with differing protein ratios throughout pregnancy. J Physiol Pharmacol 63:225–239

Nazifi S, Razavi SM, Safi N et al (2012) Malignant ovine theileriosis: alterations in the levels of homocysteine, thyroid hormones and serum trace elements. J Bacteriol Parasitol 3:1–4

Neubauer V, Griesmaier E, Ralser E, Kiechl-Kohlendorfer U (2012) The effect of sex on outcome of preterm infants—a population-based survey. Acta Paediatr 101:906–911

Neufeld KM, Kang N, Bienenstock J et al (2011) Reduced anxiety-like behavior and central neurochemical change in germ-free mice. Neurogastroenterol Motil 23:255–264

Nguon K, Ladd B, Baxter MG et al (2004) Sexual dimorphism in cerebellar structure, function, and response to environmental perturbation. Prog Brain Res 148:343–351

Nguon K, Baxter MG, Sajdel-Sulkowska EM (2005) Perinatal exposure to polychlorinated biphenyl (PCBs) differentially affects cerebellar and neurocognitive functions in male and female rat neonates. Cerebellum 4:1–11

Nguyen TT, Mol KA, DiStefano JJ III (2004) Thyroid hormone production rates in rat liver and intestine in vivo, a novel graph theory and experimental solution. Am J Physiol Endocr Metab 285:E171–E181

Nguyen LY, Ramanathan M, Weinstock-Guttman B (2009) Sex differences in vitro proinflammatory cytokine production from peripheral blood of multiple sclerosis patients. J Neurol Sci 2003:93–99

Nikoleishvili EG, Nemsadze KP, Sidamonidze LG (2005) Functional state of hypophysial-thyroid system of newborns with intrauterine infection. Georgian Med News 123:31–35

Oyhenart EE, Orden B, Fucini MC et al (2003) Sexual dimorphism and postnatal growth of intrauterine growth retarded rats. Growth Dev Aging 67:73–83

Passos MC, Lisboa PC, Pereira-Toste F et al (2012) Developmental plasticity in thyroid function primed by maternal hyperleptinemia in early lactation: a time-course study in rats. Horm Metab Res 44:520–526

Patole S (2007) Prevention and treatment of necrotising enterocolitis in preterm neonates. Early Hum Dev 83:635–642

Patterson PH (2002) Maternal infection: window on neuroimmune interactions in fetal brain development and mental illness. Curr Opin Neurobiol 12:115–118

Pearce BD (2001) Schizophrenia and viral infection during development; a focus on mechanisms. Mol Psychiatry 6:634–646

Pereira L, Petitt M, Fong A (2014) Intrauterine growth restriction caused by underlying congenital cytomegalovirus infection. J Infect Dis 209(10):1573–1584. doi:10.1093/infdis/jiv019

Raikos V, Dassios T (2014) Health-promoting properties of bioactive peptides derived from milk proteins in infant food: a review. Dairy Sci Technol 94:91–101

Rohl C, Armbrust E, Herbst E et al (2010) Mechanisms involved in the modulation of astroglial resistance to oxidative stress induced by activated microglia; antioxidative systems, peroxide elimination, radical generation, lipid peroxidation. Neurotox Res 17:317–331

Romano T, Wark JD, Owens JA et al (2009) Prenatal growth restriction and postnatal growth restriction followed by accelerated growth independently program reduced bone growth and strength. Bone 45:132–141

Rutter M, Caspi A, Moffitt T (2003) Using sex differences in psychopathology to study mechanisms: unifying issues and research strategies. J Child Psychol Psychiatry 44:1092–1115

Sajdel-Sulkowska EM, Koibuchi N (2011) Sexually-dimorphic nature of PCB effects on the developing animal CNS. In: Kestler L, Lewis M (eds) Gender differences in effects of prenatal substance exposure. The American Physiological Association Books, Washington

Sajdel-Sulkowska EM, Zabielski R (2013) Chapter 4: Gut microbiome and brain-gut axis in autism—aberrant development of gut-brain communication and reward circuitry. In: Fitzgerald M (ed) Recent advances in autism spectrum disorders, vol 1. Open Access, Rijeka. doi:10.5772/55425. ISBN 978-953-51-1021-7

Sajdel-Sulkowska EM, Bialy M, Cudnoch-Jedrzejewska A (2015) Chapter 4: Abnormal brain BDNF, "leaky gut" and altered microbiota in autism. In: Bennet C (ed) Brain derived neurotrophic factor. Nova Science, Hauppauge, pp 147–180. ISBN 978-1-63483-761-3

Sangwan N, Sangwan AK, Singh S et al (2003) Cortisol and thyroid hormones in relation to bovine tropical theileriosis. Indian J Anim Sci 72:1098–1099

Scarpignato C, Pelosini I (1999) Management of irritable bowel syndrome: novel approaches to the pharmacology of gut motility. Can J Gastroenterol 13:50A–65A

Schendel DE (2002) Infection in pregnancy and cerebral palsy. J Am Med Womens Assoc 56:105–108

Schendel DE, Schuchat A, Thorsen P (2001) Public health issues related to infection in pregnancy and cerebral pulsy. Ment Retard Dev Disabil Res Rev 8:39–45

Scott FJ, Baron-Cohen S, Bolton P, Brayne C (2002) Brief report; prevalence of autism spectrum conditions in children aged 5-11 years in Cambridgeshire, UK. Autism 6:231–237

Sebai H, Gadacha W, Sani M (2009) Protective effect of resveratrol against lipopolysaccharide-induced oxidative stress in rat brain. Brain Inj 23:1089–1094

Sharangpani A, Takanohashi A, Bell MJ (2008) Caspase activation in fetal rat brain following experimental intrauterine inflammation. Brain Res 1200:138–145

Sharma S, Dabla PK, Kumar S (2013) Thyroid hormone dysfunction and CRP levels in neonates with sepsis. J Endocr Metab 3:62–66

Silva JE (1983) Role of circulating thyroid hormones and local mechanisms in determining the concentration of T3 in various tissues. Prog Clin Biol Res 116:23–44

Sinha A, Madden J, Ross-Degnan D et al (2003) Reduced risk of neonatal. Pediatrics 112:e303

Stolp HB, Dziegielewska KM, Ek CJ et al (2005) Long-term changes in blood-brain barrier permeability and white matter following prolonged systemic inflammation in early development in the rat. Eur J Neurosci 22:2805–2816

Sulkowski ZL, Chen T, Midha S et al (2012) Perinatal exposure to thimerosal results in a sex- and strain-dependent neurodevelopmental abnormalities. Cerebellum 11:575–586

Tenore A, Parks JS, Gasparo M et al (1980) Thyroidal response to peroral TSH in suckling and weaned rats. Am J Physiol 238:E428–E430

Urakubo A, Jarskog LF, Lieberman JA et al (2001) Prenatal exposure to maternal infection alters cytokine expression in the placenta, amniotic fluid, and fetal brain. Schizophr Res 47:27–36

Uzel N, Neyzi O (1986) Thyroid function in critically ill infants with infections. Pediatr Infect Dis 5:516–519

Vain NE, Farina D, Vazquez LN (2012) Neonatology in the emerging countries: the strategies and health-economics challenges related to prevention of neonatal and infant infections. Early Hum Dev 88:S53–S59

Van Den Eaden SK, Tanner CM, Bernstein AL (2003) Incidence of Parkinson's disease: variation by age, gender, and race/ethnicity. Am J Epidemiol 157:1015–1022

van der Poll T, Endert E, Coyle SM et al (1999) Neutralization of TNF does not influence endotoxin induced changes in thyroid hormone metabolism in humans. Am J Physiol 276:R357–R362

Wang Y, Zhang Z, Ge P (2009) Iodine status and thyroid function of pregnant, lactating women and infants (0-1 yr) residing in areas with an effective Universal Salt Iodization program. Asia Pac J Clin Nutr 18:34–40

Watson JB, Mednick SA, Huttunen M et al (1999) Prenatal teratogens and the development of adult mental illness. Dev Psychopathol 11:457–466

Williams F, Delahunty C, Cheetham T (2013) Factors affecting neonatal thyroid function in preterm infants. http://dx.doi.org/10.4021/jem167w

Wlodek ME, Mibus A, Tan A (2007) Normal lactational environmental restores nephron endowment and prevents hypertension after placental restriction in the rats. J Am Soc Nephrol 18:1688–1696

Wynne O, Horvat JC, Osei-Kumah A et al (2011) Early life infection alters BALB/c hippocampal gene expression in a sex specific manner. Stress 14:247–261

Xu M, Sulkowski ZL, Parekh P et al (2013) Effect of perinatal lipopolysaccharide (LPS) exposure on the developing rat brain; modeling the effect of maternal infection on the developing human CNS. Cerebellum 12:572–586

Yang ZL, Yang LY, Huang GW et al (2003) Tri-iodothyronine supplement protects gut barrier in septic rats. World J Gastroenterol 9:347–350

Yildizdas D, Onenli-Mungan N, Yapicioglu H et al (2004) Thyroid hormone levels and their relationship to survival in children with bacterial sepsis and septic shock. J Pediatr Endocrinol Metab 17:1435–1442

Zabielski R (2007) Hormonal and neural regulation of intestinal function. Livest Sci 108:32–40

Zabielski R, Godlewski MM, Guilloteau P (2008) Control of development of gastrointestinal system in neonates. J Physiol Pharmacol 59:35–54

Chapter 5
Disruption of Feedback Regulation of Thyroid Hormone Synthesis/Secretion and Brain Development

Sumiyasu Ishii and Masanobu Yamada

Abstract The hypothalamic–pituitary–thyroid axis plays a central role in the regulation of thyroid hormone homeostasis. The hypothalamus arises from diencephalon and the hormone-producing nuclei have neuronal origin. The anterior and intermediate lobes of the pituitary gland originate from the oral ectoderm, whereas the posterior lobe is derived from the neural ectoderm. Hypothalamic TRH stimulates pituitary TSH, and TSH stimulates the production of thyroid hormone. Conversely, thyroid hormone suppresses TRH and TSH. Impairment of TRH signaling or TSH signaling results in central hypothyroidism. Central hypothyroidism is much less common compared to primary hypothyroidism. Congenital central hypothyroidism is induced by several disorders including tumors, developmental defects, and gene mutations. Patients with congenital central hypothyroidism can suffer from mental retardation, poor verbal skills, attention deficits, and motor weakness, similarly to those with congenital primary hypothyroidism. In addition to the effect of hypothyroidism, the defects in TRH signaling or TSH signaling might disturb the development of the brain.

Keywords Thyrotropin-releasing hormone • Thyrotropin • Feedback regulation • Central hypothyroidism • Brain development

5.1 Introduction

The hypothalamic–pituitary–thyroid axis plays a central role in the regulation of thyroid hormone homeostasis. The levels of serum thyroid hormones thyroxin (T4) and 3,5,3′-triiodothyroxin (T3) are tightly regulated by this system. Hypothalamic

S. Ishii (✉) • M. Yamada
Department of Medicine and Molecular Science, Gunma University Graduate School of Medicine, 3-39-15 Showa-machi, Maebashi, Gunma 371-8511, Japan
e-mail: sishii@gunma-u.ac.jp; myamada@gunma-u.ac.jp

© Springer Science+Business Media New York 2016
N. Koibuchi, P.M. Yen (eds.), *Thyroid Hormone Disruption and Neurodevelopment*,
Contemporary Clinical Neuroscience, DOI 10.1007/978-1-4939-3737-0_5

"

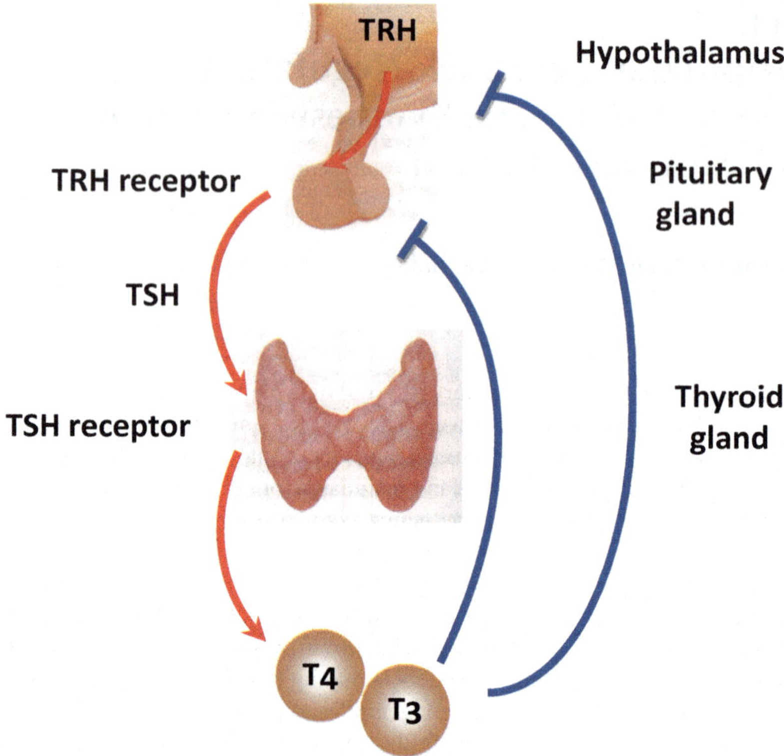

Fig. 5.1 Feedback regulation of thyroid hormone homeostasis by the hypothalamic–pituitary–thyroid axis. The hypothalamic TRH stimulates pituitary TSH, and TSH stimulates the production of thyroid hormone. Conversely the excessive thyroid hormone suppresses TRH and TSH

thyrotropin-releasing hormone (TRH) is mainly produced in the paraventricular nuclei and secreted into the hypothalamic portal vein. TRH stimulates the production and maturation of thyrotropin (thyroid-stimulating hormone, TSH) in the thyrotrophs of the anterior pituitary gland. TSH is secreted into the systemic circulation and transported to the thyroid gland, where it stimulates the synthesis and secretion of thyroid hormone. Conversely, the thyroid hormone inhibits the production of TRH and TSH (Fig. 5.1) (Braverman and Cooper 2012). This system is called the negative feedback loop. In case the serum thyroid hormone level is excessive, the levels of TRH and TSH are suppressed, and consequently, the thyroid hormone level is suppressed to normal. If the thyroid hormone level is not high enough, the production of TRH and TSH is stimulated in order to get the thyroid hormone level back to normal. The disruption of this thyroid-stimulating system results in central hypothyroidism, which is characterized by subnormal level of T4 with inappropriately low TSH level (Yamada and Mori 2008). In this chapter, we will review the role of TRH–TSH–thyroid axis from the viewpoint of neural development.

5.2 Development of the Hypothalamus and the Pituitary Gland

5.2.1 The Hypothalamus

The hypothalamus arises from diencephalon in the sixth week of gestation in human (Braverman and Cooper 2012; Yamada and Mori 2008; Moore et al. 2011; Treier and Rosenfeld 1996). The primordia of the hypothalamus appear in the ventral region on the lateral walls of the third ventricle. The epithalamus in the dorsal region and the thalamus in the middle region also develop on these walls. The hypothalamic sulcus separates the thalamus and hypothalamus, and the epithalamic sulcus lies between the epithalamus and the thalamus. The neuroblasts in the primordial hypothalamus proliferate and differentiate into hypothalamic nuclei, including the paraventricular, supraoptic, and arcuate nuclei, which produce the hormones. Each hypothalamic hormone is often produced in various nuclei, and one nucleus would produce more than one hormone. But TRH is mainly produced in the paraventricular nuclei.

5.2.2 The Pituitary Gland

The pituitary gland has two origins (Braverman and Cooper 2012; Yamada and Mori 2008; Moore et al. 2011; Treier and Rosenfeld 1996; Melmd 2010; Zhu et al. 2007). The anterior lobe and the intermediate lobe originate from the oral ectoderm. The posterior lobe is derived from the neural ectoderm. The ventral part of the hypothalamus extends inferiorly to form the infundibulum, which gives rise to the posterior pituitary gland, the median eminence, and the pituitary stalk. Several transcription factors such as Nkx2.1 (Kimura et al. 1996) and Sox3 (Rizzoti et al. 2004) play important roles in this process. Importantly, proper formation of the infundibulum is necessary for the organogenesis of the anterior pituitary gland.

In the fourth week of gestation, the middle part of the oral ectoderm invaginates and forms the hypophysial diverticulum, which is also called Rathke's pouch. Rathke's pouch extends toward the primordial hypothalamus and is separated from the oral cavity. In the eighth week, Rathke's pouch is in a closed structure and attaches the infundibulum. Many transcription factors including Pitx1/2 (Lamonerie et al. 1996; Gage et al. 1999), Lhx3/4 (Sheng et al. 1997), Hesx1 (Dattani et al. 1998), and Prop1 (Sornson et al. 1996) work coordinately to form Rathke's pouch. Subsequently the cells on the anterior wall of the pouch proliferate and start to differentiate into multiple lineages, which give rise to the pars distalis and pars tuberalis. The lumen of Rathke's pouch almost disappears. The posterior wall of the pouch becomes the pars intermedia of the pituitary gland.

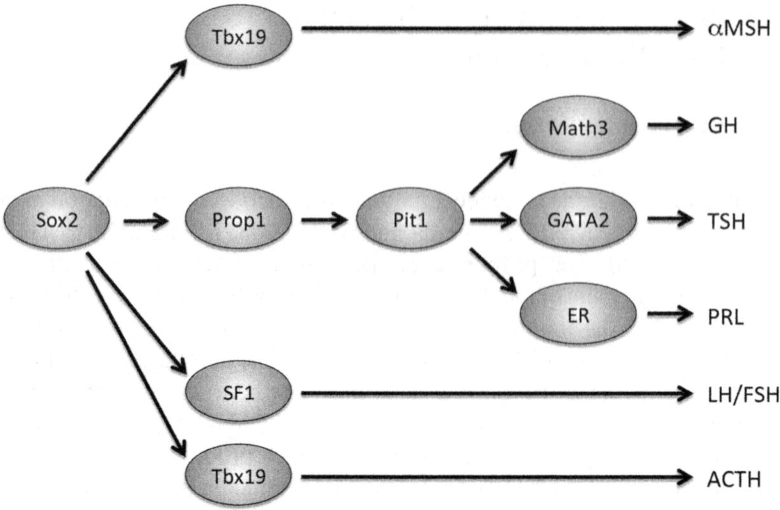

Fig. 5.2 A schematic representation of pituitary cell differentiation. The pituitary cells are differentiated from Sox2-positive putative progenitor cells. Specific transcription factors are essential for the specification of each cell type

5.2.3 Differentiation of the Anterior Pituitary Cells

The anterior lobe of the pituitary gland contains five types of hormone-producing cells. The corticotrophs produce adrenocorticotropin (ACTH), the gonadotrophs produce luteinizing hormone (LH) and follicle-stimulating hormone (FSH), the somatotrophs produce growth hormone (GH), the lactotrophs produce prolactin (PRL), and the thyrotrophs produce TSH. The melanotrophs in the intermediate lobe produce α-melanotropin (αMSH). These cell types are thought to arise from the common progenitor cells that express a transcription factor Sox2. Several transcription factors play critical roles during differentiation into each specific lineage. Prop1 is expressed in a subset of precursor cells and induces the expression of Pit1, which is a transcription factor required for the development of somatotrophs, lactotrophs, and thyrotrophs (Li et al. 1990). In the developing thyrotrophs, another transcription factor GATA2 interacts with Pit1 and induces the expression of β-subunit of TSH (Dasen et al. 1999). The commitment of other cell lineages is also regulated by specific transcription factors (Fig. 5.2).

5.2.4 The Effect of Primary Hypothyroidism on the Hypothalamic–Pituitary Development

As mentioned in other chapters, disruption of thyroid hormone affects organogenesis of the brain and other organs. But primary hypothyroidism does not affect the gross morphology of the hypothalamus and the pituitary gland. In the mass screening of newborn infants, congenital primary hypothyroidism is detected by the high level of TSH with low T4, suggesting that the negative feedback regulation is intact. However, reversible hyperplasia of the thyrotrophs is observed in patients with long-term primary hypothyroidism (Scheithauer et al. 1985).

5.3 Hormones and Receptors in the Hypothalamic–Pituitary–Thyroid Axis

5.3.1 TRH

TRH is mainly synthesized in the paraventricular nuclei of the hypothalamus. It is secreted into the hypothalamic portal vein and stimulates the production of TSH through binding to TRH receptor on the surface of thyrotrophs in the pituitary gland. TRH is essential not only for TSH production but also for the maturation of TSH. TRH regulates the conjugation of the TSH α- and β-subunit and glycosylation of the TSH molecules. Mice null for TRH gene have central hypothyroidism with slightly high serum TSH concentration, but the biological activity of TSH is decreased (Yamada et al. 1997). In addition, TRH-deficient human due to hypothalamic disorganization has TSH with low activity (Beck-Peccoz et al. 1985). TRH is also known as a stimulator of prolactin synthesis in the lactotrophs (Yamada et al. 2006).

TRH is a tripeptide pyroGlu-His-ProNH$_2$, which is produced by the cleavage of the precursor peptide preproTRH (Lechan et al. 1986). The processed progenitor TRH is further modified by convertases PC-1 and PC-2 and carboxipeptidase E and becomes mature TRH.

The synthesis of TRH is suppressed by thyroid hormone, which is a basic mechanism of the negative feedback regulation. T3 represses the expression of TRH gene at the transcriptional level via thyroid hormone receptor (Sugrue et al. 2010).

5.3.2 TRH Receptor

TRH receptor is expressed on the surface of the thyrotrophs and mediates TRH-dependent secretion of TSH. TRH receptor is a member of G-protein-coupled receptor superfamily. Upon binding to its ligand, TRH receptor activates protein kinase C pathway to stimulate the expression of α- and β-subunit of TSH as well as prolactin and other target genes (Gershengorn and Osman 1996). There is a second subtype of TRH receptor in rodents (TRH receptor 2) (Sun et al. 2003). Mice deficient in TRH receptor 1 exhibit central hypothyroidism (Rabeler et al. 2004).

5.3.3 TSH

TSH is synthesized in the thyrotrophs of the anterior pituitary gland and stimulates the production of thyroid hormone by binding to the TSH receptor that is expressed on the surface of the thyroid follicular cells. TSH is a heterodimeric glycoprotein and consists of α- and β-subunit. The α-subunit is common among three anterior pituitary hormones TSH, LH, and FSH. Chorionic gonadotropin (CG) produced in the placenta also contains the common α-subunit. The β-subunit is unique to each hormone. The transcription of the genes for α- and β-subunit of TSH is stimulated by TRH (Shupnik et al. 1986) and suppressed by T3 (Shupnik et al. 1985). After translation, two TSH subunits are glycosylated and combined, and subsequently oligosaccharides are processed (Chin et al. 1981). As mentioned above, TRH signaling plays an essential role in the maturation of TSH as well.

5.3.4 TSH Receptor

TSH binds to its receptor on the surface of the thyroid gland and stimulates the synthesis of thyroid hormone. TSH receptor is a G-protein-coupled receptor with large N-terminal extracellular domain, which binds the ligand (Smits et al. 2003). Binding of TSH activates adenylate cyclase pathway (Laurent et al. 1987) and phosphatidylinositol pathway (Kosugi et al. 1992). These pathways enhance the transcription of key molecule genes for thyroid hormone synthesis including sodium/iodine symporter, thyroid peroxidase, and thyroglobulin, thereby stimulating thyroid hormone production.

5.3.5 Thyroid Hormone and Thyroid Hormone Receptor

Thyroid hormone systemically exerts various effects including the regulation of brain development via thyroid hormone receptor. Thyroid hormone receptor belongs to nuclear receptor superfamily and controls the expression of the target genes (see Chap. 1). Thyroid hormone receptor is necessary for the negative regulation of TRH gene and TSH gene. Mutations in the thyroid hormone receptor beta gene result in resistance to thyroid hormone syndrome (Refetoff and Dumitrescu 2007). The serum TSH level of the patients with this syndrome is not suppressed despite the high level of thyroid hormones, indicating that the negative feedback loop is disturbed. Some of these patients might have defective development of the brain (see Chap. 17).

5.4   Epidemiology of Congenital Central Hypothyroidism

Hypothyroidism due to the defects in the thyroid gland is called primary hypothyroidism. On the other hand, central hypothyroidism results from the disturbance of thyroid stimulation system (Yamada and Mori 2008). Hypothyroidism induced by the defect in the pituitary gland is also called secondary hypothyroidism, whereas hypothyroidism with hypothalamic origin is tertiary hypothyroidism. The precise prevalence of congenital central hypothyroidism is unknown. It is estimated that congenital hypothyroidism is observed in 1 in 2500 newborn infants, and most of them suffer from primary hypothyroidism (Braverman and Cooper 2012). One report indicated the incidence of congenital central hypothyroidism to be 1 per 16,404 neonates (van Tijn et al. 2005).

5.5   Etiology of Congenital Central Hypothyroidism

The potential causes of central hypothyroidism are listed in Table 5.1. Hypothyroidism is either permanent or transient. Some diseases are congenital and others are acquired, and both are possible in some disorders.

5.5.1   Pituitary Adenoma

Pituitary adenomas are the most frequent reason among adult central hypothyroidism cases (including both congenital cases and acquired cases). In a Spanish study, 61 % of the adult cases were due to pituitary adenomas (Regal et al. 2001). Most of

Table 5.1   Possible causes of central hypothyroidism

	Congenital	Acquired
Permanent hypothyroidism		
Compressive lesions (tumors, developmental defects)	Yes	Yes
Vascular diseases (apoplexy, Sheehan syndrome)	Yes	Yes
Genetic mutations	Yes	No
Injuries (trauma)	No	Yes
Iatrogenic factors (surgery, drugs, radiation)	No	Yes
Transient hypothyroidism		
Gestational hyperthyroidism	Yes	No
Immunologic diseases (lymphocytic hypophysitis, sarcoidosis)	No	Yes
Infectious diseases (tuberculosis, syphilis)	No	Yes
Infiltrative lesions (histiocytosis X)	No	Yes

them would be acquired cases, but some of them could be congenital cases. Mechanical compression of the portal vessels and the pituitary stalk is postulated to be a reason for central hypothyroidism. This compression might induce ischemia of the anterior pituitary gland or insufficient delivery of hypothalamic hormones (Arafah et al. 2000). Pituitary tumor apoplexy can deteriorate the compression. Among hypopituitarism patients due to pituitary compression, the frequency of hormone deficiency is GH > LH/FSH > TSH > ACTH > PRL. It is common that multiple hormones are affected, but isolated TSH deficiency is also observed. TSH deficiencies are identified in 60 % of pituitary adenoma patients (Yamada and Mori 2008).

5.5.2 Abnormalities Related to Pituitary Development

Craniopharyngioma is a common parasellar tumor that arises from the remnant of Rathke's pouch. The remnant is observed in 13–22 % of the random autopsy cases, and the patients remain asymptomatic in most cases. However, if the tumor develops and grows, pituitary compression similar to that in pituitary tumor cases occurs. Central hypothyroidism is seen in 7–35 % of symptomatic patients (Yamada and Mori 2008).

Primary empty sella syndrome is characterized by sella turcica filled with cerebrospinal fluid and flattened pituitary gland because of raised pressure. This syndrome is observed in 6–20 % of the random autopsy cases. Twenty-eight percent of the patients exhibited hypopituitarism, and some of them had secondary hypothyroidism (Guitelman et al. 2013).

5.5.3 Gestational Hyperthyroidism

Transient thyrotoxicosis is sometimes observed in neonates born to mothers with uncontrolled Grave's disease. This is due to anti-TSH receptor antibody transferred to embryos through the placenta. In contrast, central hypothyroidism has been reported in the babies of mothers with severe thyrotoxicosis (Kempers et al. 2003). The mechanism of this phenomenon remains unclear.

5.5.4 Genetic Mutations

Mutations in several genes, including *TSHB*, *TRHR*, *POU1F1*, *PROP1*, *SOX3*, *HESX1*, *LHX3/4*, and *LEPR*, are reported to induce congenital central hypothyroidism.

TSHB gene mutation is a frequent reason for inheritable isolated central hypothyroidism. The first case was reported in 1989 and subsequently several other mutations were reported (Miyai 2007). The patients exhibit severe isolated central hypothyroidism with neonatal onset. TSH levels are low or normal and the levels of α-subunit are high. This disease is autosomal recessive.

So far central hypothyroidism induced by *TRH* gene mutation has not been reported. On the other hand, disturbed TRH signaling due to *TRHR* gene mutation results in autosomal recessive central hypothyroidism (Collu et al. 1997). The patients have relatively mild isolated central hypothyroidism. Although serum TSH level was normal, the patients exhibited blunted TSH and PRL response to TRH administration.

Mutations in the genes necessary for the development of thyrotrophs are reported as the causes of congenital central hypothyroidism. As previously mentioned, Pit1 plays a central role in the differentiation of thyrotrophs, somatotrophs, and lacto-trophs (Li et al. 1990). Pit1 is the product of *POU1F1* gene, and the mutations in this gene cause combined pituitary hormone deficiencies with severe growth retardation and distinctive facial appearance (Tatsumi et al. 1992). In addition to central hypothyroidism, these patients are defective in the production of GH and PRL. The inheritance studies show the autosomal dominant or recessive pattern depending on the mutation.

Prop1 is important for the lineage commitment of anterior pituitary cells (Sornson et al. 1996). This transcription factor is also known as an inducer of Pit1 expression. Mutations in this gene are responsible for autosomal recessive combined pituitary hormone defects, with deficiencies of TSH, GH, PRL, LH/FSH, and ACTH (Wu et al. 1998). The symptoms are various from moderate to severe, depending on the patients.

Mutations in the genes essential for early development of the pituitary gland induce combined pituitary hormone deficiencies associated with other malformations. *Sox3* (Rizzoti et al. 2004), *HESX1* (Dattani et al. 1998), *LHX3* (Netchine et al. 2000), and *LHX4* (Machinis et al. 2001) are the members of these genes.

Leptin is a hormone produced in the adipose tissue and negatively regulates appetite in the hypothalamus. Patients with disturbed leptin signaling due to the mutations in leptin receptor *LEPR* gene exhibit variable presence of central hypothyroidism combined with LH/FSH defects and severe obesity secondary to hyperphagia (Clément et al. 1998). These facts indicate that leptin signaling is an important physiologic regulator of endocrine function.

5.6 Neuronal Symptoms in Congenital Central Hypothyroidism

Hypothyroid status severely affects the neurodevelopment. The patients with congenital hypothyroidism frequently suffer from mental retardation, poor verbal skills, attention deficits, and motor weakness (see other chapters). Proper functional development is disturbed in multiple areas in the brain including the hippocampus and cerebellum (Koromilas et al. 2010; Anderson 2008). These defects are due to the impairment of myelination, axon guidance, cell migration, neurotransmission, mitochondria function, and so on.

In general, hypothyroid status is mild to moderate in patients with central hypothyroidism as compared to those with primary hypothyroidism. It is estimated that this is due to the constitutive activity of the TSH receptor (Neumann et al. 2010). However, mental retardation, cognitive defects, and motor disability are also observed in severe cases of congenital central hypothyroidism (Miyai 2007). Levothyroxine replacement therapy within first 14 days could largely restore the neurodevelopmental symptoms, but subtle impairments in mental function might be irreversible (Zoeller and Rovet 2004). The effect of mild congenital central hypothyroidism on brain development remains elusive. Studies using sensitive methods would be desired.

5.7 Experimental Approaches Using Animal Models

As tools to extensively study the pathophysiology of congenital central hypothyroidism, several animal models were generated.

TRH knockout mice have tertiary hypothyroidism with slightly high serum TSH concentration, but the biological activity of TSH is decreased (Yamada et al. 1997). Consistent with these observations, TRH-deficient human due to hypothalamic disorganization has TSH with low activity (Beck-Peccoz et al. 1985).

Rodents have two subtypes of TRH receptors (TRH receptor 1 and TRH receptor 2), but gene corresponding to *TRHR2* is not identified in human (Sun et al. 2003). There are two lines of TRH receptor 1-deficient mice reported (Rabeler et al. 2004; Zeng et al. 2007). The mice showed mild hypothyroidism but the serum TSH levels were normal (Rabeler et al. 2004). In contrast, TRH receptor 2 knockout mice are euthyroid (Sun et al. 2009), indicating the difference in the roles of two subtypes in rodents.

Behavioral studies were done in one line of TRH receptor 1-null mice and in TRH receptor 2 knockout mice, which suggest the roles of TRH signaling in brain development. TRH receptor 1 knockout mice display increased depression and anxiety-like behavior (Zeng et al. 2007). TRH receptor 2 knockout mice exhibit increased depression and reduced anxiety phenotypes (Sun et al. 2009). Although it remains unclear why the results of anxiety tests were different, these reports show the involvement of TRH signaling in the mood phenotype, which might be a result of developmental defect.

It is noteworthy that TRH receptor 2-deficient mice are euthyroid. These results indicate that TRH signaling regulates mood independently of thyroid status. In general, it is difficult to clearly distinguish the role of TRH/TSH deficiencies and that of subsequent hypothyroidism. For this purpose, T4 replacement in TRH-null mice or TRH receptor 1-null mice would be useful.

So far TSH-deficient mice are not reported. TSH receptor knockout mice show the importance of TSH signaling in the thyroid gland and the skeletal muscle

(Marians et al. 2002; Abe et al. 2003), but the role of TSH signaling in brain development has not been reported.

5.8 Potential Roles of TRH in the Brain Besides the Regulation of Thyroid Hormone Homeostasis

As discussed in the previous paragraph, TRH exerts its effects in the brain independently of thyroid hormone status. Indeed, TRH and TRH receptor are widely distributed throughout the central nervous system besides the hypothalamus or pituitary gland (Shibusawa et al. 2008). In addition, TRH has been reported to ameliorate depression (Kastin et al. 1972) and cerebellar ataxia (Sobue et al. 1980). Therefore, it would not be surprising if TRH signaling itself plays a role in brain development.

TRH stimulates the induction of cyclic GMP in the cerebellum (Mailman et al. 1978). This was the first report of TRH action outside the pituitary gland. On the other hand, NOS and cGMP contribute to cerebellar development induced by thyroid hormone (Serfozo et al. 2009). These reports suggest that TRH might play a role in developing the cerebellum by inducing cGMP independently of thyroid hormone. A cdc2-related kinase, PFTAIRE kinase 1, is proposed as a mediator of TRH–cGMP pathway (Hashida et al. 2002).

5.9 Perspectives

As compared to congenital primary hypothyroidism, much less is known about congenital central hypothyroidism, probably due to the low prevalence and difficulties in the diagnosis. Precise studies regarding brain development are still on the way. One of the interesting questions in this field would be whether developmental defect is due to the disturbed TRH/TSH signaling itself or because of subsequent hypothyroidism. Early supplementation with levothyroxine would be useful to reveal this point, either in human or in animal models. In addition, precise evaluation of the symptoms using sensitive neuronal or psychiatric tests is desired.

References

Abe E, Marians RC, Yu W et al (2003) TSH is a negative regulator of skeletal remodeling. Cell 115(2):151–162

Anderson GW (2008) Thyroid hormone and cerebellar development. Cerebellum 7(1):60–74

Arafah BM, Prunty D, Ybarra J et al (2000) The dominant role of increased intrasellar pressure in the pathogenesis of hypopituitarism, hyperprolactinemia, and headaches in patients with pituitary adenomas. J Clin Endocrinol Metab 85(5):1789–1793

Beck-Peccoz P, Amr S, Menezes-Ferreira MM et al (1985) Decreased receptor binding of biologically inactive thyrotropin in central hypothyroidism. Effect of treatment with thyrotropin-releasing hormone. N Engl J Med 312(17):1085–1090

Braverman LE, Cooper D (2012) The thyroid, 10th edn. LWW, Philadelphia

Chin WW, Maloof F, Habener JF (1981) Thyroid-stimulating hormone biosynthesis. Cellular processing, assembly, and release of subunits. J Biol Chem 256(6):3059–3066

Clément K, Vaisse C, Lahlou N et al (1998) A mutation in the human leptin receptor gene causes obesity and pituitary dysfunction. Nature 392(6674):398–401

Collu R, Tang J, Castagné J et al (1997) A novel mechanism for isolated central hypothyroidism: inactivating mutations in the thyrotropin-releasing hormone receptor gene. J Clin Endocrinol Metab 82(5):1561–1565

Dasen JS, O'Connell SM, Flynn SE et al (1999) Reciprocal interactions of Pit1 and GATA2 mediate signaling gradient-induced determination of pituitary cell types. Cell 97(5):587–598

Dattani MT, Martinez-Barbera JP, Thomas PQ et al (1998) Mutations in the homeobox gene HESX1/Hesx1 associated with septo-optic dysplasia in human and mouse. Nat Genet 19(2):125–133

Gage PJ, Suh H, Camper SA (1999) The bicoid-related Pitx gene family in development. Mamm Genome 10(2):197–200

Gershengorn MC, Osman R (1996) Molecular and cellular biology of thyrotropin-releasing hormone receptors. Physiol Rev 76(1):175–191

Guitelman M, Garcia Basavilbaso N, Vitale M et al (2013) Primary empty sella (PES): a review of 175 cases. Pituitary 16(2):270–274

Hashida T, Yamada M, Hashimoto K et al (2002) A novel TRH-PFTAIRE protein kinase 1 pathway in the cerebellum: subtractive hybridization analysis of TRH-deficient mice. Endocrinology 143(7):2808–2811

Kastin AJ, Ehrensing RH, Schalch DS et al (1972) Improvement in mental depression with decreased thyrotropin response after administration of thyrotropin-releasing hormone. Lancet 2(7780):740–742

Kempers MJ, van Tijn DA, van Trotsenburg AS et al (2003) Central congenital hypothyroidism due to gestational hyperthyroidism: detection where prevention failed. J Clin Endocrinol Metab 88(12):5851–5857

Kimura S, Hara Y, Pineau T et al (1996) The T/ebp null mouse: thyroid-specific enhancer-binding protein is essential for the organogenesis of the thyroid, lung, ventral forebrain, and pituitary. Genes Dev 10(1):60–69

Koromilas C, Liapi C, Schulpis KH et al (2010) Structural and functional alterations in the hippocampus due to hypothyroidism. Metab Brain Dis 25(3):339–354

Kosugi S, Okajima F, Ban T et al (1992) Mutation of alanine 623 in the third cytoplasmic loop of the rat thyrotropin (TSH) receptor results in a loss in the phosphoinositide but not cAMP signal induced by TSH and receptor autoantibodies. J Biol Chem 267(34):24153–24156

Lamonerie T, Tremblay JJ, Lanctôt C et al (1996) Ptx1, a bicoid-related homeo box transcription factor involved in transcription of the pro-opiomelanocortin gene. Genes Dev 10(10):1284–1295

Laurent E, Mockel J, Van Sande J et al (1987) Dual activation by thyrotropin of the phospholipase C and cyclic AMP cascades in human thyroid. Mol Cell Endocrinol 52(3):273–278

Lechan RM, Wu P, Jackson IM et al (1986) Thyrotropin-releasing hormone precursor: characterization in rat brain. Science 231(4734):159–161

Li S, Crenshaw EB III, Rawson EJ et al (1990) Dwarf locus mutants lacking three pituitary cell types result from mutations in the POU-domain gene pit-1. Nature 347(6293):528–533

Machinis K, Pantel J, Netchine I et al (2001) Syndromic short stature in patients with a germline mutation in the LIM homeobox LHX4. Am J Hum Genet 69(5):961–968

Mailman RB, Frye GD, Mueller RA et al (1978) Thyrotropin-releasing hormone reversal of ethanol-induced decreases in cerebellar cGMP. Nature 272(5656):832–833

Marians RC, Ng L, Blair HC et al (2002) Defining thyrotropin-dependent and -independent steps of thyroid hormone synthesis by using thyrotropin receptor-null mice. Proc Natl Acad Sci U S A 99(24):15776–15781

Melmd S (ed) (2010) The pituitary, 3rd edn. Elsevier, Philadelphia

Miyai K (2007) Congenital thyrotropin deficiency—from discovery to molecular biology, postgenome and preventive medicine. Endocr J 54(2):191–203

Moore KL, Persaud TVN, Torchia MG (2011) The developing human, 9th edn. Elsevier, Philadelphia

Netchine I, Sobrier ML, Krude H et al (2000) Mutations in LHX3 result in a new syndrome revealed by combined pituitary hormone deficiency. Nat Genet 25(2):182–186

Neumann S, Raaka BM, Gershengorn MC (2010) Constitutively active thyrotropin and thyrotropin-releasing hormone receptors and their inverse agonists. Methods Enzymol 485:147–160

Rabeler R, Mittag J, Geffers L et al (2004) Generation of thyrotropin-releasing hormone receptor 1-deficient mice as an animal model of central hypothyroidism. Mol Endocrinol 18(6):1450–1460

Refetoff S, Dumitrescu AM (2007) Syndromes of reduced sensitivity to thyroid hormone: genetic defects in hormone receptors, cell transporters and deiodination. Best Pract Res Clin Endocrinol Metab 21(2):277–305

Regal M, Páramo C, Sierra SM (2001) Prevalence and incidence of hypopituitarism in an adult Caucasian population in northwestern Spain. Clin Endocrinol (Oxf) 55(6):735–740

Rizzoti K, Brunelli S, Carmignac D et al (2004) SOX3 is required during the formation of the hypothalamo-pituitary axis. Nat Genet 36(3):247–255

Scheithauer BW, Kovacs K, Randall RV et al (1985) Pituitary gland in hypothyroidism. Histologic and immunocytologic study. Arch Pathol Lab Med 109(6):499–504

Serfozo Z, de Vente J, Elekes K (2009) Thyroid hormone level positively regulates NOS and cGMP in the developing rat cerebellum. Neuroendocrinology 89(3):337–350

Sheng HZ, Moriyama K, Yamashita T et al (1997) Multistep control of pituitary organogenesis. Science 278(5344):1809–1812

Shibusawa N, Hashimoto K, Yamada M (2008) Thyrotropin-releasing hormone (TRH) in the cerebellum. Cerebellum 7(1):84–95

Shupnik MA, Chin WW, Habener JF et al (1985) Transcriptional regulation of the thyrotropin subunit genes by thyroid hormone. J Biol Chem 260(5):2900–2903

Shupnik MA, Greenspan SL, Ridgway EC (1986) Transcriptional regulation of thyrotropin subunit genes by thyrotropin-releasing hormone and dopamine in pituitary cell culture. J Biol Chem 261(27):12675–12679

Smits G, Campillo M, Govaerts C et al (2003) Glycoprotein hormone receptors: determinants in leucine-rich repeats responsible for ligand specificity. EMBO J 22(11):2692–2703

Sobue I, Yamamoto H, Konagaya M et al (1980) Effect of thyrotropin-releasing hormone on ataxia of spinocerebellar degeneration. Lancet 1(8165):418–419

Sornson MW, Wu W, Dasen JS (1996) Pituitary lineage determination by the Prophet of Pit-1 homeodomain factor defective in Ames dwarfism. Nature 384(6607):327–333

Sugrue ML, Vella KR, Morales C et al (2010) The thyrotropin-releasing hormone gene is regulated by thyroid hormone at the level of transcription in vivo. Endocrinology 151(2):793–801

Sun Y, Lu X, Gershengorn MC (2003) Thyrotropin-releasing hormone receptors—similarities and differences. J Mol Endocrinol 30(2):87–97

Sun Y, Zupan B, Raaka BM et al (2009) TRH-receptor-type-2-deficient mice are euthyroid and exhibit increased depression and reduced anxiety phenotypes. Neuropsychopharmacology 34(6):1601–1608

Tatsumi K, Miyai K, Notomi T et al (1992) Cretinism with combined hormone deficiency caused by a mutation in the PIT1 gene. Nat Genet 1(1):56–58

Treier M, Rosenfeld MG (1996) The hypothalamic-pituitary axis: co-development of two organs. Curr Opin Cell Biol 8(6):833–843

van Tijn DA, de Vijlder JJ, Verbeeten B Jr et al (2005) Neonatal detection of congenital hypothyroidism of central origin. J Clin Endocrinol Metab 90(6):3350–3359

Wu W, Cogan JD, Pfäffle RW et al (1998) Mutations in PROP1 cause familial combined pituitary hormone deficiency. Nat Genet 18(2):147–149

Yamada M, Mori M (2008) Mechanisms related to the pathophysiology and management of central hypothyroidism. Nat Clin Pract Endocrinol Metab 4(12):683–694

Yamada M, Saga Y, Shibusawa N et al (1997) Tertiary hypothyroidism and hyperglycemia in mice with targeted disruption of the thyrotropin-releasing hormone gene. Proc Natl Acad Sci U S A 94(20):10862–10867

Yamada M, Shibusawa N, Ishii S et al (2006) Prolactin secretion in mice with thyrotropin-releasing hormone deficiency. Endocrinology 147(5):2591–2596

Zeng H, Schimpf BA, Rohde AD et al (2007) Thyrotropin-releasing hormone receptor 1-deficient mice display increased depression and anxiety-like behavior. Mol Endocrinol 21(11):2795–2804

Zhu X, Wang J, Ju BG et al (2007) Signaling and epigenetic regulation of pituitary development. Curr Opin Cell Biol 19(6):605–611

Zoeller RT, Rovet J (2004) Timing of thyroid hormone action in the developing brain: clinical observations and experimental findings. J Neuroendocrinol 16(10):809–818

Part II

Animal Models to Study Thyroid Hormone Disruption on Neurodevelopment

Chapter 6
Animal Models to Study Thyroid Hormone Action in Neurodevelopment

Noriaki Shimokawa and Noriyuki Koibuchi

Abstract This chapter introduces representative animal models currently used to study thyroid hormone-mediated development and functional maintenance of target organs. Particularly, the potential animal model for the research on thyroid hormone system in neurodevelopment is discussed. Several representatives are (1) congenital hypothyroid animals caused by thyroid dysgenesis or thyroid dyshormonogenesis, (2) thyroid hormone receptor (TR)-modified animals, and (3) thyroid hormone transport or metabolism-modified animals. TR is a nuclear hormone receptor, which acts as a ligand-regulated transcription factor. On thyroid hormone response element, liganded TR activates transcription of target gene, whereas unliganded TR represses the transcription. Thus, phenotype of TR knockout mouse is different from that of hypothyroid animal (low thyroid hormone level), in which unliganded TR actively represses the transcription. On the other hand, a human patient harboring mutant TR expresses a different phenotype depending on the function of mutated TR. To mimic this phenotype, knock-in mice harboring various mutated TR have been generated. In addition, recent human studies have shown that thyroid hormone transporters such as monocarboxylate transporter (MCT) 8 may play an important role in thyroid hormone-mediated brain development. However, MCT8 knockout mouse shows a different phenotype from that of a human patient. Although it is impossible to introduce the whole sets of model animal in thyroid and neurodevelopment research, this chapter may be helpful to understand an outline of this field.

Keywords Thyroid hormone • Nuclear receptor • Transcription • Development • Cretinism • Thyroid hormone resistance

N. Shimokawa, Ph.D. • N. Koibuchi, M.D., Ph.D. (✉)
Department of Integrative Physiology, Gunma University Graduate School of Medicine,
3-39-22 Showa-machi, Maebashi, Gunma 371-8511, Japan
e-mail: nkoibuch@gunma-u.ac.jp

© Springer Science+Business Media New York 2016
N. Koibuchi, P.M. Yen (eds.), *Thyroid Hormone Disruption and Neurodevelopment*,
Contemporary Clinical Neuroscience, DOI 10.1007/978-1-4939-3737-0_6

6.1 Animal Models for Thyroid Dysgenesis or Thyroid Dyshormonogenesis

There are several key factors that are essential for thyroid morphogenesis. Mutation of genes encoding such factor may produce congenital hypothyroidism due to thyroid dysgenesis or thyroid dyshormonogenesis (Park and Chatterjee 2005). The former includes thyroid transcription factors TTF-1, TTF-2, Pax8, and thyroid-stimulating hormone (TSH) receptor; the latter includes thyroid peroxidase (TPO), dual oxidase (Duox), thyroglobulin (Tg), sodium–iodide symporter, and pendrin. Representative animal models to study the effect of congenital hypothyroidism induced by such gene defect or malfunction are listed in Table 6.1.

6.1.1 Animal Models for Thyroid Dysgenesis

The Pax8 knockout mouse is one of the most commonly used hypothyroid animal models. Pax8 is essential for differentiation of thyroid follicular epithelial cell, and thus its knockout induces a severe hypothyroidism (Mansouri et al. 1998).

Table 6.1 Mutant animals showing congenital hypothyroid phenotype

Species	Name	References	Etiology	Representative phenotypes
Thyroid dysgenesis				
Mouse	*Pax8⁻/⁻*	Mansouri et al. (1998)	Pax8 gene knockout	Severe thyroid gland dysgenesis
Mouse	*hyt/hyt*	Biesiada et al. (1996)	TSH receptor mutation	Relatively milder hypothyroid phenotype
Thyroid dyshormonogenesis				
Mouse/rat	PTU or MMI treated	Legrand (1967)	Inhibition of TPO activity	Hypothyroidism at various severities
Mouse	*tpo*	Takabayashi et al. (2006)	Mutation of TPO gene	Severe hypothyroidism with goiter. Brain phenotype not known
Mouse	*thyd*	Johnson et al. (2007)	Mutation of dual oxidase 2 (Duox2) gene	Severe hypothyroidism with goiter. Hearing impairment
Mouse	*cog/cog*	Beamer et al. (1987)	Thyroglobulin gene mutation	Large goiter but mild hypothyroid phenotype
Rat	*rdw*	Umezu et al. (1998)	Thyroglobulin gene mutation	Severe hypothyroidism with thyroid gland atrophy. Various brain phenotypes
		Shimokawa et al. (2014)		Retardation of cerebellar morphogenesis
				Hypoactivity

Development and gene expression profile in the brain are greatly affected in this mouse (Poguet et al. 2003). Since the disruption of organogenesis by Pax8 knockout is seen only in the thyroid, this mouse can be a good model to study the molecular mechanisms of thyroid hormone action in the brain.

Since TSH plays a critical role for the development and function of the thyroid gland, a mutant mouse harboring mutated TSH receptor induces thyroid dysgenesis. The *hyt/hyt* mouse harbors a point mutation of C to T at nucleotide position 1666, which replaces Pro with Leu at position 556 in transmembrane domain IV of the TSH receptor (Biesiada et al. 1996). Mutation of this region may inhibit transloca-tion of TSH receptor from the cytoplasm to cell membrane, resulting in low TSH binding (Yen 2000). This mouse shows significantly delayed somatic and behav-ioral development, which is a typical phenotype for congenital hypothyroidism (Biesiada et al. 1996). However, catch-up growth occurs after weaning, and the size becomes normal in adulthood (Li and Chow 1994), and the brain phenotype is lim-ited. However, more than 20 families harboring TSH receptor mutation have been reported, and several families harbor mutation of transmembrane domain IV (Yen 2000). Thus, this mouse can be a good model to study the pathophysiology of such disease.

There are other animal models of thyroid dysgenesis, such as TTF-1 or TTF-2 knockout mice. To study the molecular mechanisms of thyroid action in the brain, however, these mouse models may not be ideal because they show defects of other organs.

6.1.2 *Animal Models for Thyroid Dyshormonogenesis*

Disruption of several critical steps of thyroid hormone synthetic pathway may induce thyroid dyshormonogenesis. For example, TPO regulates oxidation and organification of iodine and coupling of iodotyrosine to iodothyronine. Any disrup-tion of such steps may induce a decrease in thyroid hormone synthesis (Lever et al. 1983).

Antithyroid Drug-Induced Hypothyroidism

Antithyroid drugs such as propylthiouracil (PTU) and methimazole (methylmercap-toimidazole, MMI) have been used commonly to study thyroid hormone action in the brain since 1960 (Legrand 1967). These drugs inhibit synthesis of thyroid hor-mone by inhibiting TPO activity. Since perinatal administration of these drugs through drinking water can easily produce congenital hypothyroid animals, these drugs are still used commonly (Koibuchi 2001). The brain phenotype of drug-induced hypothyroidism is more severe when drugs are administered during devel-opment, compared with those of adult mice treated with such drugs, since thyroid hormone affects growth and differentiation of the brain mainly during critical developmental period. In the rodent cerebellum, for example, the effect of the anti-thyroid drugs is greatest during the first postnatal two weeks, which is a period when dendritic growth of Purkinje cell, proliferation and migration of granule cells, and synaptogenesis of cerebellar neurons are active (Koibuchi et al. 2003).

Mutant Animals Showing Thyroid Dyshormonogenesis

Thyroid dyshormonogenesis has been shown to result from a number of different genes. Many mutant mouse models harboring mutation of genes responsible for thyroid hormone synthetic pathway have been reported (Grasberger and Refetoff 2011). To study the role of thyroid hormone in the brain, however, abnormal mor-phogenesis that is directly induced by a particular gene mutation should be confined within the thyroid gland. Animal models that may fit for such criteria are those harboring TPO (Takabayashi et al. 2006; Johnson et al. 2007; Zamproni et al. 2008) or Tg gene mutation (Beamer et al. 1987; Umezu et al. 1998; Shimokawa et al. 2014). The advantage of using these mice over PTU- or MMI-treated animals is that, since dams are essentially euthyroid, indirect effects to pup development pro-duced by maternal behavioral alteration can be excluded.

Takabayashi et al. (2006) have reported a natural dwarf mutant mouse, named *tpo*, showing a severe hypothyroid phenotype (a distinct growth retardation with short life span) with reduced T3 and T4 and elevated TSH plasma levels. Although the brain phenotype has not yet been studied, *tpo* mouse is potentially a good model to study thyroid hormone action in the brain. Another mouse model with disrupted TPO activity is *thyd* mouse, in which gene encoding dual oxidase 2 (Duox2) is mutated (Johnson et al. 2007). Since Duox produces hydrogen peroxide that is essential for the action of TPO in organification of iodide, this mutation causes a severe hypothyroidism. Although its brain phenotype has not yet been fully exam-ined, this mouse shows a hearing impairment that is a typical neurological pheno-type of congenital hypothyroidism. Thus, this mouse is also potentially a good model to study thyroid hormone action in the brain. In addition to Duox2 gene mutation, another report has shown that mutation of Duox maturation factor 2 (Duoxa2), which is required to express Duox2 enzymatic activity, causes congenital hypothyroidism in human (Zamproni et al. 2008). The animal models harboring Duoxa mutation have recently been generated (Grasberger et al. 2012). Although the brain phenotype has not yet been examined, this mouse could be an interesting animal model to study.

Another group of animals showing congenital hypothyroid phenotype is those harboring mutations of Tg gene. Thyroid hormone is made within this molecule. Tg protein consists of 2749 and 2766 amino acids in human and mouse, respectively, and many mutations have been identified from many species (Rivolta and Targovnik 2006). In mouse, a congenital hypothyroid mice strain with a large goiter induced by mutation of Tg gene is known as *cog/cog* mouse (Beamer et al. 1987). Although the detailed analysis has not yet been done, its cerebellar weight is significantly less than those of normal mice (Sugisaki et al. 1992). However, hypomyelination is restricted in the cerebrum, and its general growth returns to normal with the advance of age, indicating that its hypothyroidism is rather mild. On the other hand, a rat congenital hypothyroid strain named as *rdw* rat shows a severe hypothyroidism with no goiter formation (Umezu et al. 1998). Recently, we have reported a severe impairment of motor coordination and balance (Shimokawa et al. 2014). This phenotype may be due to retardation of cerebellar morphogenesis, which correlates with the small somata and poor dendritic arborization of Purkinje cells and retarded migration of granule cells during the first two postnatal weeks. Moreover, the rats showed hypoactivity due to insufficient axonal transport of dopamine from the substantia nigra to the striatum. Since not many hypothyroid rat models are available, this rat is useful to study brain function and behavioral disorders in congenital hypothyroidism.

6.2 Animal Models for Thyroid Hormone Receptor Gene Mutation

Many research groups have generated different kinds of TR gene-modified mice, particularly in the late 1990s to the early 2000s. Representable models are shown as follows.

6.2.1 TR Knockout Mice

In the presence of T3, TR activates transcription of target gene harboring thyroid hormone response element, whereas, without T3, TR actively suppresses transcription. Thus, the effect of TR deletion is different from those of thyroid dyshormonogenesis, since the repressor activity of unliganded TR is deleted in TR-null mice. To study the role of TR on organ development and function, however, TR knockout mice are essential. There are two TR genetic loci termed as α and β, each of which produces several functional TRs as a result of alternative splicing and/or differential promoter usage. Furthermore, some introns have a weak promoter activity such as intron 7 of TRα gene. Thus, deletion of upstream exon may result in the expression of additional TR-related proteins, whose expression may be limited under normal condition (Chassande 2003). So far, at least three additional TR-related proteins may be generated. Such proteins, termed as TR$\Delta\alpha$1, TR$\Delta\alpha$2, and TR$\Delta\beta$3, lack N-terminus and DNA-binding domain (DBD) and cannot bind to TRE. Thus, phenotypes of TR knockout mice may be due to combination of deletion of a specific TR with overexpression of other TR species. Table 6.2 shows the list of TR knockout mice. Possible remaining TR proteins in each animal are also indicated.

TRα1-deleted mice show slightly reduced thyroid function with almost normal phenotype except for 20 % reduced heart rate, prolonged QRS and QT durations,

Table 6.2 Thyroid hormone receptor (TR) knockout mice

Gene	Targeted exon	References	Deleted TRs	Remained TRs	Representative phenotypes
TRα					
TRα1$^{-/-}$	Exon 9	Wikström et al. (1998)	α1, Δα1	α2, Δα2, all β	Normal T3 with slightly reduced T4 level
		Guadaño-Ferraz et al. (2003)			Prolonged QRS and QT durations
		Morte et al. (2002, 2004)			Prevention of hypothyroid phenotype in the cerebellum
		Peeters et al. (2013)			
TRα2$^{-/-}$	Exon 10	Saltó et al. (2001)	α2, Δα2	α1, Δα1, all β	Overexpression of TRα1, inducing both hyper- (high body temperature, increased hear rate) and hypothyroid phenotypes (increased body fat)
TRα$^{-/-}$	Exon 2	Fraichard et al. (1997)	α1, α2	Δα1, Δα2, all β	Aberrant intestine and bone development
TRα$^{0/0}$	Exon 5–intron 7	Gauthier et al. (2001)	All α	All β	Aberrant intestine and bone development, but the phenotype is less severethanthoseinTRα$^{-/-}$
		Macchia et al. (2001)			
TRβ					
TRβ2$^{-/-}$	Exon 2	Abel et al. (1999)	β2	β1 (β3, Δβ3)	Central resistance to thyroid hormone levels
		Ng et al. (2001)		All α	Elevated TSH, T3, and T4
					Selective loss of M-cone in the retina
TRβ$^{-/-}$	Exon 3	Forrest et al. (1996)	All β	All α	Central resistance to thyroid hormone
		Sandhofer et al. (1998)			Elevated TSH, T3, and T4 levels
					Aberrant auditory function development
TRα and *TRβ*					

(continued)

Table 6.2 (continued)

Gene	Targeted exon	References	Deleted TRs	Remained TRs	Representative phenotypes
TRα1$^{-/-}$TRβ$^{-/-}$	See above	Göthe et al. (1999)	α1, Δα1, all β	α2, Δα2	High T3 and T4 levels due to high TSH
					Growth retardation. Abnormal bone maturation
TRα$^{-/-}$ TRβ$^{-/-}$	TRα$^{-/-}$: see above	Gauthier et al. (1999)	α1, α2, all β	Δα1, Δα2	Aberrant intestine/ bone development (more severe than TRα$^{-/-}$)
	TRβ$^{-/-}$: exon 4–5				Elevated TSH,T3, and T4 levels (more severe than TRβ$^{-/-}$)
TRα$^{0/0}$ TRβ$^{-/-}$	TRα$^{0/0}$: see above	Gauthier et al. (2001)	All α	None	Reduced body temperature and bone maturation (more severe than TRα$^{-/-}$)
	TRβ$^{-/-}$: exon 4–5		All β		Aberrant auditory function (more severe than TRβ$^{-/-}$)
					Aberrant intestine development (milder than TRα$^{-/-}$ or TRα$^{-/-}$ TRβ$^{-/-}$)

and 0.5 °C lower body temperature (Wikström et al. 1998). A limited alteration of behavior and neural circuit is also reported (Guadaño-Ferraz et al. 2003). However, their cerebellar phenotype appeared to be normal except for aberrant maturation of astrocytes (Morte et al. 2004). Another group also showed that the cerebellar morphology in TRα1-deleted mice (TRα$^{-/-}$) is not overtly different from that in TRα$^{+/+}$ mice (Peeters et al. 2013). More strikingly, deletion of TRα1 prevents structural alteration of the cerebellum in hypothyroidism that is induced by MMI and perchlorate treatment (Morte et al. 2002). These results indicate that abnormal cerebellar phenotype in thyroid dyshormonogenetic animal may be due to dominant-negative action of unliganded TRα proteins. On the other hand, TRα2 knockout mouse shows both hyper- and hypothyroid phenotype in an organ-specific manner (Saltó et al. 2001). This may be due to elevated expression of TRα1 in this mouse. TRα1 expression in the brain is also elevated, but cerebellar phenotype is not clear. Deletion of both TRα1 and TRα2 also shows only limited phenotype in the cerebellum. However, apart from cerebellar phenotype, the existence of TRΔα1 and/or TRΔα2 shows altered phenotype in various organs. When TRα1 and TRα2 are deleted but TRΔα1 and TRΔα2 expressions are not inhibited (TRα$^{-/-}$) (Fraichard et al. 1997), their phenotype is more severe than those of mice in which all TRα proteins are

deleted (TRα$^{0/0}$) (Gauthier et al. 2001; Macchia et al. 2001). The greater decrease in plasma thyroid hormone levels, the more severe impairment of bone and intestine development is observed.

More limited brain phenotype is observed in TRβ knockout mice. While TRβ1 is widely expressed in the brain, the expression of TRβ2 is confined within the pituitary, hypothalamus (thyrotropin-releasing hormone (TRH) neuron), retina, and inner ear (Koibuchi et al. 2003). TRβ2 knockout mice show central resistance to thyroid hormone with elevated T3, T4, and TSH levels in serum (Abel et al. 1999). Furthermore, this deletion causes a selective loss of M-cones in the retina (Ng et al. 2001). However, abnormal brain phenotype seems to be confined within the hypothalamus, and any change in cerebellar and cerebral phenotype has not been reported. In TRβ knockout mouse, on the other hand, it shows aberrant development of auditory function in addition to central hypothyroidism (Forrest et al. 1996). However, although TRβ is strongly expressed in the brain such as in the Purkinje cell, its deletion does not induce any alteration of thyroid hormone-responsive genes in the cerebellum (Sandhofer et al. 1998).

In TRα and TRβ double-knockout mice, one receptor cannot substitute the function for the other. Thus, their phenotypes are more severe than those of single-gene knockout. In TRα1$^{-/-}$TRβ$^{-/-}$ mice, delayed general growth and aberrant bone maturation are observed, which are not seen in each single-knockout mouse (Göthe et al. 1999). In TRα$^{-/-}$TRβ$^{-/-}$ mice, aberrant intestinal development, which is seen in TRα1$^{-/-}$, and high T3, T4, and TSH levels, which are seen in TRβ$^{-/-}$, are observed; both of which are more severe than those of single-knockout mice (Gauthier et al. 1999). However, in TRα$^{0/0}$TRβ$^{-/-}$ mice, while low body temperature and abnormal auditory function, which are more severe than those of TRα$^{0/0}$ or TRβ$^{-/-}$, respectively, are seen, aberrant intestinal development is milder than those of TRα$^{-/-}$TRβ$^{-/-}$ or TRα1$^{-/-}$ (Gauthier et al. 2001). These results indicate the possible contribution of TRα variants (Δα1 and/or Δα2) in generating differential phenotypes.

6.2.2 TR Knock-in Mice

The syndromes of resistance to thyroid hormone (RTH) are characterized as reduced action of thyroid hormone in thyroid hormone target tissue (Refetoff et al. 1993). The majority of cases of RTH is generalized RTH, whereas pituitary RTH is also reported. In addition to elevated serum levels of T3 and T4 levels with non-suppressed TSH, many patients show clinical features related to neurological disorders such as hyperactivity and learning disability. Animal models that mimic the RTH patients are generated (Table 6.3). The majority of patients harbor a mutation in the TRβ gene. Several knock-in mice harboring mutated human TRβ have been generated (Kaneshige et al. 2000; Hashimoto et al. 2001). Their phenotypes are similar to those seen in human RTH patient such as elevated levels of T3 and T4 with unsuppressed TSH levels in serum, delayed general growth, and goiter. At least one of these mutations shows an aberrant cerebellar development similar to those

Table 6.3 Mutant TR knock-in mice

Gene	Replaced TR	References	Representative phenotypes
TRβ1	*PV	Kaneshige et al. (2000)	Growth retardation. Decrease in thyroid hormone-sensitive gene expression in various organ
			Elevated T3, T4, and TSH levels
	D337T	Hashimoto et al. (2001)	Abnormal development of the cerebellum and hippocampus
		Portella et al. (2010)	Elevated T3, T4, and TSH levels
TRα1	*PV	Kaneshige et al. (2001)	Growth retardation. Increase in mortality rate
		Itoh et al. (2001)	Mild elevation of T3, T4, and TSH levels
			Decrease in cerebellar gene expression
	R384C	Tinnikov et al. (2002)	Growth retardation. Severe neurological abnormalities including impaired locomotor activity possibly due to impaired cerebellar development
		Venero et al. (2005)	
	L400R	Quignodon et al. (2007)	Growth retardation. Delayed cerebellum development
		Fauquier et al. (2011)	Decrease in cerebellar gene expression
		Avci et al. (2012)	Aberrant developmental loss of axonal regenerative capacity

*PV, PV are initials of a patient harboring a mutation in exon 10 of TRβ gene, a C-insertion at codon 448, which produces a frameshift of the carboxy-terminal 14 amino acids of TRβ1

seen in congenital hypothyroid animals (Hashimoto et al. 2001). These TRβ-mutant mice, carrying a natural human mutation (Δ 337 T) in the TRβ locus, show decreased arborization of Purkinje cell dendrite with aberrant locomotor activity and decreased expression of thyroid hormone-responsive genes in the cerebellum (Hashimoto et al. 2001). Portella et al. (2010) reported that these mice showed smaller cerebellum area characterized by impaired lamination and foliation, severe deficits in proliferation of granular precursors, arborization of Purkinje cells, and organization of Bergmann glia fibers. In addition, although RTH patient harboring mutated TRα gene has not yet been identified, knock-in mice harboring mutant TRα have also been generated to examine the involvement of unliganded TRα on development and functional maintenance of target organ including the cerebellum (Kaneshige et al. 2001; Itoh et al. 2001; Tinnikov et al. 2002; Venero et al. 2005; Quignodon et al. 2007; Fauquier et al. 2011; Avci et al. 2012). Compared to TRβ-mutated mice, their neurological phenotype is more severe. This tendency is evident when the phenotypes of mice harboring the same point mutation in TRα and TRβ are compared (Itoh et al. 2001). Mutant TRα knock-in mice show various aberrant cerebellar developments similar to those seen in congenital hypothyroid animals. These results indicate that TRα may be more involved in regulating cerebellar gene expression than TRβ.

Quignodon et al. (2007) generated new transgenic mice that express activation function 2 (AF-2) mutation (L400R) in TRα1. This mutation prevents the recruitment of histone acetyl transferase (HAT) and sustains the recruitment of corepressors permanently. As a result, therefore, this mutation can be responsible for dominant-negative activity in thyroid hormone signaling. The mutant TRα1 knock-in mice also show various aberrant phenotypes in the cerebellum (Quignodon et al. 2007; Fauquier et al. 2011): delay of cerebellar development, reduction of the number and size of dendritic spines, persistence of the external granular layer (EGL), abnormal Bergmann glia maturation, and downregulation of neurotrophic factor such as NT3 and BDNF. Using this mouse, Avci et al. (2012) have reported that the loss of axonal regenerative capacity in Purkinje cells induced by physiological T3 burst during development is mainly mediated by TRα1 and involves in its downstream target molecule Krüppel-like factor 9. These results indicate that TRα1 has a critical role in a fundamental process of postnatal maturation of the brain by thyroid hormone signaling. Thus, this animal model contributed greatly to understand further the molecular mechanisms of thyroid hormone action in neurodevelopment.

6.3 Thyroid Hormone Transport or Metabolism-Modified Animals

6.3.1 Thyroid Hormone Transporter-Modified Animals

Circulating thyroid hormone cannot cross blood–brain barrier (BBB) without specific transporters (Visser et al. 2008). At least two transporters play a role in transporting thyroid hormone across BBB. One is monocarboxylate transporter (MCT) 8; the other is organic anion-transporting polypeptide (Oatp) 1c1. Transthyretin (TTR), a major thyroid hormone-binding protein particularly in rodent, is also expressed at the choroid plexus and has been considered to play an important role in transporting thyroid hormone across the choroid plexus–cerebrospinal fluid barrier. However, thyroid hormone concentration in the brain is not altered in TTR knockout mice, suggesting that TTR is not required for thyroid hormone distribution in the brain (Palha et al. 2002). After entering the brain, thyroid hormone is taken up by astrocyte or tanycyte in which T4 is converted to T3, an active form of thyroid hormone, by type 2 iodothyronine deiodinase (DI) and transferred to neuron or oligodendrocyte (Koibuchi et al. 2003). Thyroid hormone is further deiodinated by type 3 DI, which is mainly expressed in neurons. Modification of these thyroid hormone metabolic pathways may affect greatly the development and function of the brain including the cerebellum. Model mice harboring deletion of these genes are summarized in Table 6.4.

Recently, Allan–Herndon–Dudley syndrome, a novel syndrome that is associated with a severe mental retardation, low muscle tone, spasticity with episodic involuntary movement, the absence of speech development, and elevated T3 levels

Table 6.4 Gene-modified mice harboring deletion of gene involved in thyroid hormone metabolic pathway

Gene	References	Representative phenotypes
Transporters		
MCT8$^{-/-}$	Dumitrescu et al. (2006)	Elevated T3 and TSH and decreased T4 in serum. Decreased T3 and T4 levels in the brain. Much milder neurological phenotype than those of patient harboring the same mutation
	Trajkovic et al. (2007)	
MCT8$^{-/-}$TRHR1$^{-/-}$	Trajkovic-Arsic et al. (2010)	Upregulated TRH expression in the periventricular hypothalamic nucleus
		Decreased TSH in the pituitary. Decreased T3 and T4 levels in serum
MCT8$^{-/-}$Pax8$^{-/-}$	Trajkovic-Arsic et al. (2010)	Athyroidism. Death after weaning without thyroid hormone injection
MCT10$^{-/-}$	Müller et al. (2014)	Normal serum and tissue thyroid hormone levels
MCT10$^{-/-}$MCT8$^{-/-}$	Müller et al. (2014)	Normal serum T4 levels. Partial rescue of Mct8 knockout phenotype
Oatp1c1$^{-/-}$	Mayerl et al. (2012)	Decreased T3 and T4 levels in the brain
Oatp1c1$^{-/-}$MCT8$^{-/-}$	Mayerl et al. (2014)	Locomotor abnormalities due to delayed cerebellar development and reduced myelination. Altered deiodinase activities and thyroid hormone target gene expression
Deiodinase		
Dio2$^{-/-}$	Schneider et al. (2001)	Decrease in T3 content in the brain. Milder neurological phenotype than those of hypothyroid animal
	Galton et al. (2007)	
Dio3$^{-/-}$	Hernandez et al. (2006)	Marked elevation of T3 during perinatal development, inducing upregulation of cerebellar thyroid hormone-responsive genes
Dio3$^{-/-}$TRα$^{-/-}$	Peeters et al. (2013)	Marked improvement in cerebellar morphology and the thickness of the external granule cell layer compared with Dio3-deficient mice

in serum, is reported. These patients harbor mutated MCT8 (Dumitrescu et al. 2004). Thus, this protein may play a critical role in thyroid hormone-mediated brain development. Recently, two separate groups generated MCT8 knockout mice (Dumitrescu et al. 2006; Trajkovic et al. 2007). Although the pattern of serum thyroid hormone levels is similar to those seen in a human patient, histological examination shows no abnormal development in the brain including the Purkinje cell, which expresses MCT8, despite the low levels of thyroid hormone in the brain (Trajkovic et al. 2007). The cause of the discrepancy of phenotype between human and mouse model has not yet been elucidated. Further study may be required to clarify the cause of such discrepancy. In MCT8 and TRH receptor 1 double-knockout

(Mct8$^{-/-}$Trhr1$^{-/-}$) mice, the activity of the hypothalamic–pituitary–thyroid (HPT) axis in the mice has been reported (Trajkovic-Arsic et al. 2010). Furthermore, MCT8 and Pax8 double knockout (Mct8$^{-/-}$Pax8$^{-/-}$) are completely athyroid and die after weaning without the injection of thyroid hormone (Trajkovic-Arsic et al. 2010). More recently, analysis of MCT10 knockout mice revealed normal serum thyroid hormone levels and tissue thyroid hormone content in contrast to these of Mct8 knockout mice (Müller et al. 2014). Mct10 and Pax8 double knockout (Mct10$^{-/-}$Mct8$^{-/-}$) showed normal serum T4, brain T4 content, and hypothalamic TRH expression, indicating that the additional inactivation of MCT10 partially rescues the phenotype by Mct8 gene defect (Müller et al. 2014).

In addition to MCT8 and MCT10, Oatp1c1 may also play an important role in thyroid hormone-mediated brain development. Oatp1c1-deficient mice develop without any overt neurological abnormalities despite the low levels of thyroid hormone in the brain (Mayerl et al. 2012). On the other hand, uptake of thyroid hormones into the CNS of Oatp1c1/MCT8 double-knockout mice is strongly reduced (Mayerl et al. 2014). Consistent with poor arborization and reduced dendritic growth of Purkinje cells, MCT8/Oatp1c1 DKO mice showed pronounced defect of motor coordination and reduced locomotor activities. It seems that these new transporter gene-deficient mice are extremely useful to the study of thyroid hormone action in the brain.

6.3.2 Iodothyronine Deiodinase-Modified Animals

A type 2 DI-deleted mouse (Dio2$^{-/-}$) has been generated (Schneider et al. 2001). This mouse shows decreased T3 content in most brain regions including the cerebellum, which is as severe as those in hypothyroid mouse brain. However, the change in thyroid hormone-responsive genes in the cerebellum and behavioral alteration such as motor coordination defect is much milder than those in hypothyroid mice (Galton et al. 2007). These results indicate that a possible compensate mechanism may play a role to minimize functional abnormalities. Knockout mouse for type 3 DI also has neurological phenotype (Hernandez et al. 2006). Since this enzyme is involved in inactivation of T3, deletion of this gene induces a marked elevated level of T3 during perinatal development, inducing an upregulation of thyroid hormone-responsive genes such as in the cerebellum. Type 3 DI-deficient mouse (Dio3$^{-/-}$) displayed reduced foliation, accelerated disappearance of the EGL, and premature expansion of the molecular layer of the cerebellum (Peeters et al. 2013). On the other hand, type 3 DI and TRα1 double-knockout (Dio3$^{-/-}$TRα$^{-/-}$) mice showed a marked improvement in cerebellar morphology and the thickness of the EGL compared with Dio3$^{-/-}$ mice (Peeters et al. 2013). These results indicate that type 3 DI need to protect cerebellar tissues from premature stimulation by thyroid hormone at around perinatal period, because the enzyme is expressed in the mouse cerebellum at embryonic and neonatal stages. It should be noted that there is a type 1 DI that is mainly expressed in the liver. Deletion of type 1 DI does not alter brain development (Schneider et al. 2006).

6.4 Animals Showing Similar Cerebellar Phenotype as Hypothyroid Animals

Several additional animal models show neurological phenotypes similar to those seen in hypothyroid animals. For example, a natural mutant mouse called *staggerer*, which harbors a mutated retinoid receptor-related orphan receptor (ROR) α, shows cerebellar phenotypes similar to those in hypothyroid animal, although their thyroid function is normal. Such similarity is partly induced through the cross talk between RORα and TR; thyroid hormone regulates RORα gene expression during postnatal cerebellar development (Koibuchi and Chin 1998), and RORα is required for full function of TR-mediated transcription (Qiu et al. 2007). Since TR and RORα are coexpressed in many neurons including the Purkinje cell, disruption of RORα function may result in the hypothyroid-like phenotype due to such cross talk.

Another animal model, steroid receptor coactivator-1 (SRC-1) knockout mouse, shows a similar cerebellar phenotype as congenital hypothyroid animals. Since cofactor proteins are required for TR function and SRC-1 is most abundantly expressed in the cerebellum (Martinez de Arrieta et al. 2000), deletion of SRC-1 gene also induces hypothyroid-like cerebellar phenotype (Nishihara et al. 2003). Indeed, phenotypes of such animals are induced not only by disruption of thyroid hormone system but also by other pathways. However, these mice may be also useful to analyze the mechanisms of TR-mediated transcription.

6.5 Conclusion

In this chapter, animal models that may be useful to study the role of thyroid hormone and/or TR in neurodevelopment have been described. These are classified as (1) congenital hypothyroid animals due to thyroid gland dysgenesis or thyroid dyshormonogenesis, (2) TR gene-mutated animals, and (3) thyroid hormone transport or metabolism-modified animals. Interestingly, not all such animal models show hypothyroid phenotype. While thyroid dysgenetic or thyroid dyshormonogenetic animals show a typical congenital hypothyroid brain phenotype, TR knockout animals show relatively mild or no brain phenotype. This may be because TR has bidirectional functions. While it activates transcription of target genes in the presence of ligand, it actively represses transcription in the absence of ligand. Thus, hypothyroid phenotype in "low thyroid hormone" animals may be induced through unliganded TR, whereas TR knockout animals show milder phenotype due to the absence of active repression. This hypothesis is confirmed further by the congenital hypothyroid animal-like cerebellar phenotypes in animals expressing mutated TR that does not bind to thyroid hormones. By comparing the several animal models, therefore, the physiological roles of thyroid hormone in the brain can be studied further.

Thyroid hormone regulates the expression of most target genes in the brain only at limited critical period during neurodevelopment. The molecular mechanisms generating such critical period have not yet been clarified. Because the development of rodent cerebellum occurs mostly postnatal and because anatomical structure of the cerebellum is well characterized, the cerebellum can be a very good system to study such mechanisms. However, cerebellar phenotypes of thyroid hormone system-related animal models have not always been examined, probably because some researchers who generated animals may not be interested in the brain. Further studies are thus required to examine the change in cerebellar structure and gene expression in animal models discussed in this chapter.

References

Abel ED, Boers ME, Pazos-Moura C et al (1999) Divergent roles for thyroid hormone receptor beta isoforms in the endocrine axis and auditory system. J Clin Invest 104:291–300

Avci HX, Lebrun C, Wehrlé R et al (2012) Thyroid hormone triggers the developmental loss of axonal regenerative capacity via thyroid hormone receptor α1 and krüppel-like factor 9 in Purkinje cells. Proc Natl Acad Sci U S A 109:14206–14211

Beamer WG, Maltais LJ, DeBaets MH et al (1987) Inherited congenital goiter in mice. Endocrinology 120:838–840

Biesiada E, Adams PM, Shanklin DR et al (1996) Biology of the congenitally hypothyroid *hyt/hyt* mouse. Adv Neuroimmunol 16:309–346

Chassande O (2003) Do unliganded thyroid hormone receptors have physiological functions? J Mol Endocrinol 31:9–20

Dumitrescu AM, Liao XH, Best TB et al (2004) A novel syndrome combining thyroid and neurological abnormalities is associated with mutations in a monocarboxylate transporter gene. Am J Hum Genet 74:168–175

Dumitrescu AM, Liao XH, Weiss RE et al (2006) Tissue-specific thyroid hormone deprivation and excess in monocarboxylate transporter (mct) 8-deficient mice. Endocrinology 147:4036–4043

Fauquier T, Romero E, Picou F et al (2011) Severe impairment of cerebellum development in mice expressing a dominant-negative mutation inactivating thyroid hormone receptor alpha1 isoform. Dev Biol 356:350–358

Forrest D, Erway LC, Ng L et al (1996) Thyroid hormone receptor beta is essential for development of auditory function. Nat Genet 13:354–357

Fraichard A, Chassande O, Plateroti M et al (1997) The T3R alpha gene encoding a thyroid hormone receptor is essential for post-natal development and thyroid hormone production. EMBO J 16:4412–4420

Galton VA, Wood ET, St Germain EA et al (2007) Thyroid hormone homeostasis and action in the type 2 deiodinase-deficient rodent brain during development. Endocrinology 148:3080–3088

Gauthier K, Chassande O, Plateroti M et al (1999) Different functions for the thyroid hormone receptors TRα and TRβ in the control of thyroid hormone production and post-natal development. EMBO J 18:623–631

Gauthier K, Plateroti M, Harvey CB et al (2001) Genetic analysis reveals different functions for the products of the thyroid hormone receptor alpha locus. Mol Cell Biol 21:4748–4760

Göthe S, Wang Z, Ng L et al (1999) Mice devoid of all known thyroid hormone receptors are viable but exhibit disorders of the pituitary-thyroid axis, growth, and bone maturation. Genes Dev 13:1329–1341

Grasberger H, Refetoff S (2011) Genetic causes of congenital hypothyroidism due to dyshormonogenesis. Curr Opin Pediatr 23:421–428

Grasberger H, De Deken X, Mayo OB et al (2012) Mice deficient in dual oxidase maturation factors are severely hypothyroid. Mol Endocrinol 26:481–492

Guadaño-Ferraz A, Benavides-Piccione R, Venero C et al (2003) Lack of thyroid hormone receptor alpha1 is associated with selective alterations in behavior and hippocampal circuits. Mol Psychiatry 8:30–38

Hashimoto K, Curty FH, Borges PP et al (2001) An unliganded thyroid hormone receptor causes severe neurological dysfunction. Proc Natl Acad Sci U S A 98:3998–4003

Hernandez A, Martinez ME, Fiering S et al (2006) Type 3 deiodinase is critical for the maturation and function of the thyroid axis. J Clin Invest 116:476–484

Itoh Y, Esaki T, Kaneshige M et al (2001) Brain glucose utilization in mice with a targeted mutation in the thyroid hormone α or β receptor gene. Proc Natl Acad Sci U S A 98:9913–9918

Johnson KR, Marden CC, Ward-Bailey P et al (2007) Congenital hypothyroidism, dwarfism, and hearing impairment caused by a missense mutation in the mouse dual oxidase 2 gene, Duox2. Mol Endocrinol 21:1593–1602

Kaneshige M, Kaneshige K, Zhu X et al (2000) Mice with a targeted mutation in the thyroid hormone beta receptor gene exhibit impaired growth and resistance to thyroid hormone. Proc Natl Acad Sci U S A 97:13209–13214

Kaneshige M, Suzuki H, Kaneshige K et al (2001) A targeted dominant negative mutation of the thyroid hormone α1 receptor causes increased mortality, infertility, and dwarfism in mice. Proc Natl Acad Sci U S A 98:15095–15100

Koibuchi N (2001) Thyroid hormone and cerebellar development. In: Manto M, Pandolfo M (eds) The cerebellum and its disorders. Cambridge University Press, Cambridge, pp 305–315

Koibuchi N, Chin WW (1998) ROR alpha gene expression in the perinatal rat cerebellum: ontogeny and thyroid hormone regulation. Endocrinology 139:2335–2341

Koibuchi N, Jingu H, Iwasaki T et al (2003) Current perspectives on the role of thyroid hormone in growth and development of cerebellum. Cerebellum 2:279–289

Legrand J (1967) Variations, en fonction de l'age, de la réponse du cervelet a l'action morphogénétique de la thyroïde chez le rat. Arch Anat Microsc Morphol Exp 56:291–308

Lever EG, Medeiros-Neto GA, DeGrood LJ (1983) Inherited disorders of thyroid metabolism. Endocr Rev 4:213–239

Li J, Chow SY (1994) Subcellular distribution of carbonic anhydrase and Na$^+$-K$^+$-ATPase in the brain of the *hyt/hyt* hypothyroid mice. Neurochem Res 19:83–88

Macchia PE, Takeuchi Y, Kawai T et al (2001) Increased sensitivity to thyroid hormone in mice with complete deficiency of thyroid hormone receptor alpha. Proc Natl Acad Sci U S A 98:349–354

Mansouri A, Chowdhury K, Gruss P (1998) Follicular cells of the thyroid gland require Pax8 gene function. Nat Genet 19:87–90

Martinez de Arrieta C, Koibuchi N, Chin WW (2000) Coactivator and corepressor gene expression in rat cerebellum during postnatal development and the effect of altered thyroid status. Endocrinology 141:1693–1698

Mayerl S, Visser TJ, Darras VM et al (2012) Impact of Oatp1c1 deficiency on thyroid hormone metabolism and action in the mouse brain. Endocrinology 153:1528–1537

Mayerl S, Müller J, Bauer R et al (2014) Transporters MCT8 and OATP1C1 maintain murine brain thyroid hormone homeostasis. J Clin Invest 124:1987–1999

Morte B, Manzano J, Scanlan T et al (2002) Deletion of the thyroid hormone receptor alpha 1 prevents the structural alterations of the cerebellum induced by hypothyroidism. Proc Natl Acad Sci U S A 99:3985–3989

Morte B, Manzano J, Scanlan TS et al (2004) Aberrant maturation of astrocytes in thyroid hormone receptor alpha 1 knockout mice reveals an interplay between thyroid hormone receptor isoforms. Endocrinology 145:1386–1391

Müller J, Mayerl S, Visser TJ et al (2014) Tissue-specific alterations in thyroid hormone homeostasis in combined Mct10 and Mct8 deficiency. Endocrinology 155:315–325

Ng L, Hurley JB, Dierks B et al (2001) A thyroid hormone receptor that is required for the development of green cone photoreceptors. Nat Genet 27:94–98

Nishihara E, Yoshida-Komiya H, Chan CS et al (2003) SRC-1 null mice exhibit moderate motor dysfunction and delayed development of cerebellar Purkinje cells. J Neurosci 23:213–222

Palha JA, Nissanov J, Fernandes R et al (2002) Thyroid hormone distribution in the mouse brain: the role of transthyretin. Neuroscience 113:837–847

Park SM, Chatterjee VK (2005) Genetics of congenital hypothyroidism. J Med Genet 42:379–389

Peeters RP, Hernandez A, Ng L et al (2013) Cerebellar abnormalities in mice lacking type 3 deiodinase and partial reversal of phenotype by deletion of thyroid hormone receptor α1. Endocrinology 154:550–561

Poguet AL, Legrand C, Feng X et al (2003) Microarray analysis of knockout mice identifies cyclin D2 as a possible mediator for the action of thyroid hormone during the postnatal development of the cerebellum. Dev Biol 254:188–199

Portella AC, Carvalho F, Faustino L et al (2010) Thyroid hormone receptor beta mutation causes severe impairment of cerebellar development. Mol Cell Neurosci 44:68–77

Qiu CH, Shimokawa N, Iwasaki T et al (2007) Alteration of cerebellar neurotropin messenger ribonucleic acids and the lack of thyroid hormone receptor augmentation by staggerer-type retinoic acid receptor-related orphan receptor-alpha mutation. Endocrinology 148:1745–1753

Quignodon L, Vincent S, Winter H et al (2007) A point mutation in the activation function 2 domain of thyroid hormone receptor α1 expressed after CRE-mediated recombination partially recapitulates hypothyroidism. Mol Endocrinol 21:2350–2360

Refetoff S, Weiss RE, Usala SJ (1993) The syndromes of resistance to thyroid hormone. Endocr Rev 14:348–399

Rivolta CM, Targovnik HM (2006) Molecular advances in thyroglobulin disorders. Clin Chim Acta 374:8–24

Saltó C, Kindblom JM, Johansson C et al (2001) Ablation of TRα2 and a concomitant overexpression of alpha1 yields a mixed hypo- and hyperthyroid phenotype in mice. Mol Endocrinol 15:2115–2128

Sandhofer C, Schwartz HL, Mariash CN et al (1998) Beta receptor isoforms are not essential for thyroid hormone-dependent acceleration of PCP-2 and myelin basic protein gene expression in the developing brains of neonatal mice. Mol Cell Endocrinol 137:109–115

Schneider MJ, Fiering SN, Pallud SE et al (2001) Targeted disruption of the type 2 selenodeiodinase gene (DIO2) results in a phenotype of pituitary resistance to T4. Mol Endocrinol 15:2137–2148

Schneider MJ, Fiering SN, Thai B et al (2006) Targeted disruption of the type 1 selenodeiodinase gene (Dio1) results in marked changes in thyroid hormone economy in mice. Endocrinology 147:580–589

Shimokawa N, Yousefi B, Morioka S et al (2014) Altered cerebellum development and dopamine distribution in a rat genetic model with congenital hypothyroidism. J Neuroendocrinol 26:164–175

Sugisaki T, Beamer WG, Noguchi T (1992) Microcephalic cerebrum with hypomyelination in the congenital goiter mouse (cog). Neurochem Res 17:1037–1040

Takabayashi S, Umeki K, Yamamoto E et al (2006) A novel hypothyroid dwarfism due to the missense mutation Arg479Cys of the thyroid peroxidase gene in the mouse. Mol Endocrinol 20:2584–2590

Tinnikov A, Nordström K, Thorén P et al (2002) Retardation of postnatal development caused by a negatively acting thyroid hormone receptor α1. EMBO J 21:5079–5087

Trajkovic M, Visser TJ, Mittag J et al (2007) Abnormal thyroid hormone metabolism in mice lacking the monocarboxylate transporter 8. J Clin Invest 117:627–635

Trajkovic-Arsic M, Müller J, Darras VM et al (2010) Impact of monocarboxylate transporter-8 deficiency on the hypothalamus-pituitary-thyroid axis in mice. Endocrinology 151:5053–5062

Umezu M, Kagabu S, Jiang J et al (1998) Evaluation and characterization of congenital hypothyroidism in rdw dwarf rats. Lab Anim Sci 48:496–501

Venero C, Guadaño-Ferraz A, Herrero AI et al (2005) Anxiety, memory impairment, and locomotor dysfunction caused by a mutant thyroid hormone receptor alpha1 can be ameliorated by T3 treatment. Genes Dev 19:2152–2163

Visser WE, Friesema EC, Jansen J et al (2008) Thyroid hormone transport in and out of cells. Trends Endocrinol Metab 19:50–56

Wikström L, Johansson C, Saltó C et al (1998) Abnormal heart rate and body temperature in mice lacking thyroid hormone receptor alpha 1. EMBO J 17:455–461

Yen PM (2000) Thyrotropin receptor mutations in thyroid diseases. Rev Endocr Metab Disord 1:123–129

Zamproni I, Grasberger H, Cortinovis F et al (2008) Biallelic inactivation of the dual oxidase maturation factor 2 (DUOXA2) gene as a novel cause of congenital hypothyroidism. J Clin Endocrinol Metab 93:605–610

Chapter 7
Thyroid Hormone Receptor Mutation and Neurodevelopment

Jens Mittag

Abstract Insufficient levels of thyroid hormone during development are detrimental for the brain, and the resulting defects can last into adulthood. Likewise, mutations in the nuclear thyroid hormone receptors severely interfere with brain development. Thyroid hormone receptor $\alpha1$, which is the major receptor isoform in almost all neurons, plays the pivotal role in mediating the effects of thyroid hormone in the brain; however, some distinct functions are also governed by thyroid hormone receptor β.

On the neuroanatomical level, impaired thyroid hormone receptor signaling leads to reduced myelination and branching, decreased synaptogenesis, and problems in neuronal migration, often affecting the number of neurons in certain brain areas. Particularly sensitive to changes in thyroid hormone signaling during development are neurons containing the neurotransmitter γ-aminobutyric acid (GABA) in the cortex, hippocampus, cerebellum, or hypothalamus. The anatomical defects are accompanied by consequent changes in motoric functions, memory, spatial navigation, or alterations in endocrine and autonomic homeostasis. Although transgene animal models for mutations in thyroid hormone receptors have greatly contributed in dissecting the distinct roles of thyroid hormone receptor isoforms in brain development and function, only a few cellular and molecular targets have been thoroughly characterized to date. This chapter summarizes the current knowledge on thyroid hormone receptors in neurodevelopment.

Keywords Brain • Hypothyroidism • Resistance to thyroid hormone • Cortex • Hippocampus • Hypothalamus • Cerebellum • Parvalbumin • Neuron

J. Mittag (✉)
Center of Brain Behavior and Metabolism, University of Lübeck,
Ratzeburger Allee 160, Lübeck 23562, Germany
e-mail: jens.mittag@uni-luebeck.de

© Springer Science+Business Media New York 2016
N. Koibuchi, P.M. Yen (eds.), *Thyroid Hormone Disruption and Neurodevelopment*,
Contemporary Clinical Neuroscience, DOI 10.1007/978-1-4939-3737-0_7

7.1 Introduction

Thyroid hormone is indispensable for proper pre- and postnatal brain development, as evidenced, for example, by the severe mental retardation resulting from untreated congenital hypothyroidism (De Felice and Di Lauro 2004). Similarly, maternal hypothyroxinemia, for example due to insufficient iodine availability during pregnancy, can lead to neurological cretinism including spastic diplegia in the offspring (DeLong et al. 1985). Even mild forms of developmental hypothyroidism are associated with delays and permanent changes in the brain, as the organ is exquisitely sensitive to changes in thyroid hormone status and has only a restricted capacity for compensation (Berbel et al. 2009; de Escobar et al. 2007; Ghassabian et al. 2011; Oppenheimer et al. 1991; Sharlin et al. 2010). This plethora of clinical observations emphasizes the importance of temporally and spatially well-concerted action of the hormone for optimal brain development.

The concept has emerged that thyroid hormone acts by interfering with proliferation, thus triggering the precursor cells in the brain to develop. This has been described for neural stem cells as well as oligodendrocytes (Chen et al. 2011; Rodriguez-Pena 1999). Consequently, if thyroid hormone action occurs too early during development, the number of precursor cells is reduced, while a delayed action as in hypothyroidism leads to a later onset of development. Even a minor delay can have devastating consequences, as migratory or developmental cues by the surrounding environment might be missed, leading to permanent and irreversible defects, such as insufficient myelination, reduced branching, impaired dendritic-axonal interactions, and defective synaptogenesis (Bernal 2005; Horn and Heuer 2010). Moreover, thyroid hormone is also needed for proper angiogenesis in the developing brain (Zhang et al. 2010). Taken together, thyroid hormone exerts a delicate function in brain development, and any disturbances in availability are likely to have severe consequences that potentially last into adulthood.

7.2 Thyroid Hormone Receptors

While the clinical consequences of developmental hypothyroidism are well known, the underlying molecular mechanisms have remained incompletely understood. Several rodent models have been generated to obtain a deeper understanding; however, certain species-specific differences have complicated the interpretation. For instance, rodents are born earlier in their development compared to humans; therefore the latter are even more dependent on maternal thyroid function. Moreover, there are major differences in the repertoire of thyroid hormone transporters between the species (Horn and Heuer 2010). Nevertheless, with regard to the receptors that mediate thyroid hormone action in the brain, animal models have provided invaluable insights.

Almost all actions of thyroid hormone are mediated by its nuclear receptors (TRs). These receptors are encoded by two separate genes, thra and thrb, which are located on chromosomes 11 and 14 in the mouse and 17 and 3 in humans, respectively. The genes give rise to four major thyroid hormone receptor isoforms (Fig. 7.1), TRα1 and TRα2 as well as TRβ1 and TRβ2, and several truncated short forms with unknown function. While TRα1, TRβ1, and TRβ2 are fully functional thyroid hormone receptors with DNA and ligand binding domains, TRα2 lacks the ability to bind to the hormone. Despite high levels of expression, the role of TRα2 remains enigmatic to date (Cheng et al. 2010; Salto et al. 2001).

During brain development, the expression of TRs precedes the function of the embryonal thyroid gland. In humans, the receptors already appear in the first trimester, whereas the fetal thyroid gland only starts functioning in the second trimester (Bernal and Pekonen 1984; Iskaros et al. 2000). In rodents, the first T3 binding in the brain is detected at embryonic day 13.5 (E13.5), concurring with the appearance of the first TRα1 protein in the cortical plate and the marginal zone in the same time (Wallis et al. 2010). The fetal thyroid gland starts functioning at E17.5, around 4 days before birth (Bradley et al. 1992; Perez-Castillo et al. 1985). Consequently, this gap between receptor appearance and endogenous ligand production needs to be bridged by maternal hormone to induce TR signaling (Quignodon et al. 2004).

In the brain, around 70–80 % of thyroid hormone binding is mediated by TRα1 (Schwartz et al. 1992). The receptor is present in almost all postmitotic neurons as well as oligodendrocyte precursors and specialized glia cells called tanycytes, which line the third ventricle of the hypothalamus (Wallis et al. 2010). In contrast, TRβ appears later during development at around E15.5 (Bradley et al. 1992), with TRβ1 being initially present mainly in zones of neuroblast proliferation, and later widespread in oligodendrocytes (Carre et al. 1998; Rodriguez-Pena 1999). TRβ2 expression is more restricted and found in the developing hippocampus and striatum, and later in the paraventricular nucleus of the hypothalamus (Bradley et al. 1989, 1992; Mellstrom et al. 1991). Given the very minor overlap between TRα1 and TRβ expression in the brain, it seems unlikely that the receptors exert common functions or compensate for each other to a significant degree.

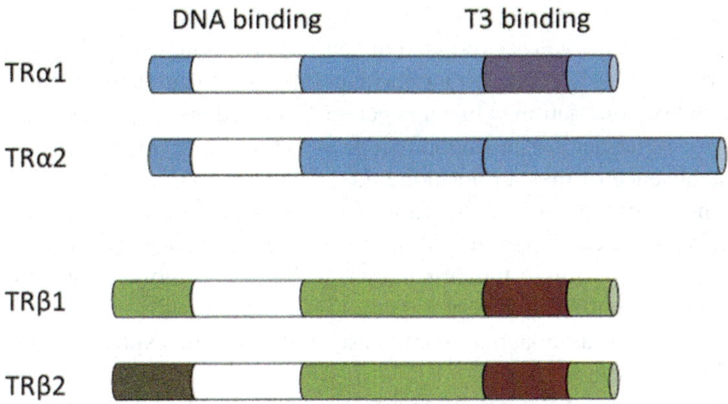

Fig. 7.1 Thyroid hormone receptors. Two distinct genes (thra and thrb) give rise to four major thyroid hormone receptor isoforms: TRα1, TRα2 and TRβ1, TRβ2. While all four isoforms can bind to DNA, only TRα1, TRβ1, and TRβ2 can bind to the ligand T3

In concurrence with the early appearance and the widespread neuronal expression, TRα1 plays a more important role for brain development and function than TRβ. Initial studies demonstrated that mice with impaired TRβ signaling have normal brain glucose utilization, while animals with defects in TRα1 show reductions in almost all investigated neuroanatomical areas (Itoh et al. 2001). Moreover, mice lacking TRβ did not show any gross abnormalities in brain morphology, suggesting only subtle and specific roles for this receptor in brain development and function (Forrest et al. 1996; O'Shea and Williams 2002). However, all results obtained from adult TR mouse models should be interpreted with caution, as they represent a mixture of developmental and acute defects. Even more importantly, mice with defective TRβ signaling display high endogenous level of thyroid hormone. It is therefore often difficult to distinguish whether a phenotypic abnormality is the result of the lack of TRβ or the high levels of hormone acting on the intact TRα1.

In summary, given the presence of TRα1 in almost every neuron and its early expression, this receptor isoform is requisite for brain development and function. TRβ is likely to have very defined roles in the brain; however, the analysis of its functions is complicated by the fact that defective TRβ signaling is usually accompanied by endogenous hyperthyroidism.

7.3 Aporeceptor Function

Thyroid hormone receptors mediate gene expression by binding to specific regions on the DNA termed thyroid hormone response elements (TREs). In contrast to many other DNA binding motifs, the sequences of TREs are highly variable, and it is usually impossible to identify functional TREs in silico. Moreover, due to the poor quality of the commercially available TR antibodies, genome-wide TRE screenings using chromatin immunoprecipitation have provided variable results. Only recently, expression systems with tagged TRs opened the road for genome-wide TRE screenings in vitro and in vivo (Chatonnet et al. 2013; Dudazy-Gralla et al. 2013). It is estimated that around 10 % of the genes expressed in neurons are regulated by thyroid hormone (Chatonnet et al. 2013). Those with a positive TRE display an increase in expression upon thyroid hormone action, while a negative TRE mediates suppression.

Like some other nuclear hormone receptors, TRs can modulate gene expression even in the absence of thyroid hormone, i.e., as aporeceptors. In case of a positive TRE, for instance, the aporeceptor (apo-TR) can bind to the response element as homodimer or as heterodimer with retinoic-X-receptor (RXR) and recruit corepressors to silence expression of the target gene. If thyroid hormone is available, it will form a complex with the receptor (holo-TR), which will in turn release the corepressors and recruit coactivators, thus stimulating gene expression (Bernal and Morte 2013; Cheng et al. 2010).

As a consequence of the potent aporeceptor activity, the lack of ligand is more severe than the lack of the TR itself. In the absence of TRs, the hormone will not be able to induce gene expression; however, the gene will still be expressed at a basal level. In the case of insufficient hormone on the other hand, as in hypothyroidism, the apo-TR will actively suppress gene expression below the basal level, causing a more severe phenotype. This is well documented when comparing the severe phenotype of congenitally hypothyroid Pax8−/− mice that lack the thyroid gland (Mansouri et al. 1998) and mice devoid of all TRs (Gothe et al. 1999): while Pax8−/− mice do not survive weaning (Flamant et al. 2002; Mittag et al. 2005), TRα1−/−TRβ−/− double mutants are viable. With regard to the brain, the defects are similarly more pronounced in hypothyroidism than in TR knockout animals. Even more remarkably, the hypothyroid phenotype can often be ameliorated when TRα1 is additionally inactivated, as this will prevent the deleterious consequences of the apo-TRα1 (Gil-Ibanez et al. 2013). During organ development however, the switch from active suppression by the apo-TR to active induction by the holo-TR is considered a physiologically relevant mechanism (Mai et al. 2004). This further emphasizes the importance of properly controlling the timing of thyroid hormone action during embryogenesis.

Knockout mice lacking all hormone binding TR isoforms are therefore inadequate for studying the consequences of developmental hypothyroidism in the brain, as they lack TR aporeceptor activity. Consequently, more recent studies have focused on mice expressing mutant TR isoforms with low or absent affinity to thyroid hormone, thereby generating a constantly repressive TR. Indeed, the phenotype of these mice more closely resembled features found in hypothyroidism (Hashimoto et al. 2001; Itoh et al. 2001; Tinnikov et al. 2002).

7.4 Brain Defects Caused by Impaired Thyroid Hormone Receptor Function

Despite the plethora of defects caused by developmental hypothyroidism, the available knowledge about the cellular and molecular targets of thyroid hormone is still limited. This is largely owed to the complexity of the brain: although a certain cell type might be identified by its expression of a specific marker protein, they nevertheless often serve distinct functions in different areas of the brain. Even within the same anatomical region, seemingly uniform cell populations can still be very heterogeneous. Therefore, any molecular work aiming to define the actions of thyroid hormone in the brain is always based on variably heterogeneous samples, depending on if whole brain extracts, punches from defined nuclei, or even sorted cells are used. Moreover, it is often impossible to assign a defined population of cells in the brain to a specific physiological or behavioral defect in the animal; often several layers of central control exist in the neuronal circuits, including self-regulatory feedback mechanisms.

Despite these difficulties, genetic mouse models have significantly aided in elucidating the actions of thyroid hormone in the brain on a cellular and molecular level. This includes animal models for congenital hypothyroidism such as the hyt/hyt or the Pax8−/− mouse (Beamer et al. 1981; Mansouri et al. 1998), mice lacking either or both forms of thyroid hormone receptor (Forrest et al. 1996; Gothe et al. 1999; Wikstrom et al. 1998), and the more recent strains expressing a mutant thyroid hormone receptor (Itoh et al. 2001; Tinnikov et al. 2002). However, the phenotype of the adult mice often reflects a combination between acute and developmental hypothyroidism, and therefore a careful analysis is required. Moreover, subtle differences between seemingly similar model systems can lead to contradicting results, which are difficult to interpret (Chatonnet et al. 2011). Consequently, newer studies have focused on conditional knockout/knockin animal models that allow a better temporospatial resolution (Picou et al. 2012), or animal models with a milder mutation that can be reactivated during defined windows of development or in adulthood, e.g., the TRα1+m mouse (Tinnikov et al. 2002). These approaches have greatly contributed to the identification of new thyroid hormone targets in the brain, and narrowed the critical developmental window of thyroid hormone sensitivity. The sections below summarize the current knowledge on the role of thyroid hormone receptors in neuronal development in specific brain areas (Fig. 7.2). A common theme will emerge showing that systems depending on the neurotransmitter GABA (gamma aminobutyric acid) are especially sensitive to developmental hypothyroidism (de Guglielmone and Gomez 1966; Wiens and Trudeau 2006).

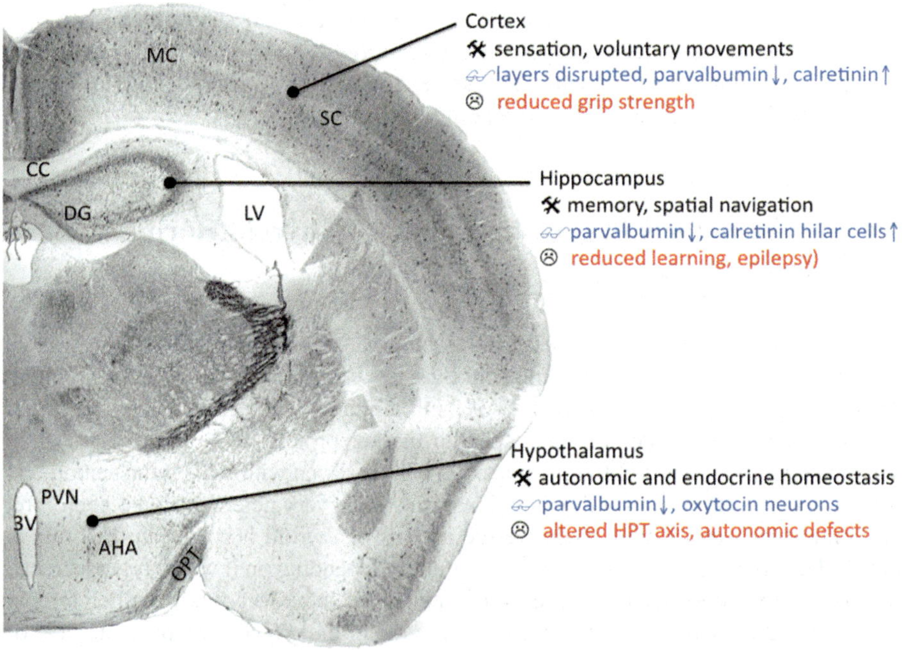

Fig. 7.2 Forebrain regions affected by developmental hypothyroidism. Mouse forebrain section stained for parvalbumin protein, illustrating three regions particularly affected by hypothyroidism during development: cortex, hippocampus, and hypothalamus. The most prominent function of the anatomical area (⚒), the anatomical defects observed upon defective TR signaling during development (〰), and the resulting phenotypical consequences (☹) are given for these regions. *3V* third ventricle of the hypothalamus, *AHA* anterior hypothalamic area, *CC* corpus callosum, *DG* dentate gyrus of the hippocampus, *LV* lateral ventricle, *MC* motor cortex, *OPT* optical tract, *PVN* paraventricular nucleus of the hypothalamus, *SC* somatosensory cortex

7.4.1 Cortex

The cortex consists of several defined layers and governs important brain functions, especially the integration of sensory information such as touch (somatosensory cortex), sight (visual cortex), and sound (auditory cortex). Furthermore, it controls voluntary movements (motor cortex). Hypothyroidism during cortical development strongly interferes with its anatomical architecture, leading to decreased number and length of radial glia, loss of neuronal bipolarity, and impaired neuronal migration (Berbel et al. 2010; Pathak et al. 2011). As a consequence of this misguided neuronal migration, populations of neurons can be found ectopically for instance in the corpus callosum (Goodman and Gilbert 2007), and the original cortical layering is perturbed (Berbel et al. 2010; Wallis et al. 2008). On the cellular level, this is most noticeable for GABAergic neurons expressing the marker parvalbumin. The emergence of these cells is severely retarded during postnatal development by hypothyroidism—a defect that eventually leads to fewer and malfunctioning cells in adulthood (Berbel et al. 1996; Gilbert and Lasley 2013; Wallis et al. 2008). This is accompanied by a reduced activity of GAD (glutamate decarboxylase), the enzyme producing the neurotransmitter GABA, in the hypothyroid cortex. Interestingly, another population of GABAergic neurons expressing calretinin is slightly increased under hypothyroid conditions (Wallis et al. 2008). However, the underlying molecular mechanisms remain unclear, e.g., whether the apparent lack of parvalbumin neurons is caused by the true absence of this cell type (due to differentiation to another cell type or misguided migration), or whether the precursors are present but unable to express parvalbumin.

Despite the obvious neuroanatomical abnormalities in the hypothyroid cortex, the consequences for animal physiology and behavior are less clear. A motoric defect, as assessed by the hanging wire test, has been associated to disturbed cortical TRα1 signaling (Wallis et al. 2008); however, this defect could not be fully reversed by postnatal treatment with thyroxine, indicating that the cellular cortical composition is not alone responsible for the motoric deficiencies associated with developmental hypothyroidism.

Similarly, the molecular mechanism underlying the cellular defects remains enigmatic. Although several target genes of thyroid hormone receptors in the developing cortex have been identified to date, such as KCNJ10, CBR2, CIRBP, or ANGPTL14 (Gil-Ibanez et al. 2013), we are far from understanding the pathways connecting thyroid hormone signaling to cortical neuron development. Some pathways initially identified that involved brain-derived neurotrophic factor (BDNF) could not be confirmed in further studies and remain controversial (Gilbert and Lasley 2013; Pathak et al. 2011).

7.4.2 *Hippocampus*

The hippocampus, located beneath the cerebral cortex, is mainly involved in memory formation and spatial navigation. It also plays a key role in epilepsy or Alzheimer's disease. Similar to what is observed in the cortex, a reduced number of parvalbumin neurons is observed if thyroid hormone is impaired during hippocampal development (Guadano-Ferraz et al. 2003; Venero et al. 2005), with the most pronounced lack of GABAergic neurons in the dentate gyrus (Gilbert et al. 2007). Surprisingly, the calretinin expressing hilar cell types are more than ten-fold increased in the dorsal regions of the hippocampus (Hadjab-Lallemend et al. 2010). Despite these pronounced developmental defects, treatment with thyroxine can still partially restore the reduced number of parvalbumin nerve terminals in some parts of the hippocampus in the adult animal (Venero et al. 2005), suggesting that $TR\alpha1$ signaling is still important for hippocampal function in adulthood. However, some disturbances for instance in excitatory postsynaptic potentials cannot be ameliorated, as they require intact thyroid hormone signaling during E10 to E13 (Liu et al. 2010; Wang et al. 2012).

In contrast to cortical functions, hippocampal integrity can be more easily assessed by behavioral studies in rodents. Tests to evaluate spatial learning, such as the Morris water maze, revealed severe memory defects as a consequence of impaired thyroid hormone signaling during development, for instance in the congenitally hypothyroid hyt/hyt mice or in rats exposed to maternal hypothyroidism during development (Anthony et al. 1993; Liu et al. 2010; Wang et al. 2012). Similar to the neuroanatomical defects, some of these phenotypical impairments are the consequence of acutely disturbed $TR\alpha1$ signaling: $TR\alpha1+m$ mice display cognitive impairments in the Novel Object Recognition test and increased anxiety behavior in the open field or the elevated plus maze; however, these abnormalities are largely reversed by thyroxine treatment of the adult animal, which restores acute $TR\alpha1$ signaling (Venero et al. 2005). Interestingly, an involvement of $TR\beta$ has also been discussed for hippocampal function. While mice expressing the $TR\beta\Delta377T$ mutation display abnormal hippocampal gene expression and exhibit learning problems in the Morris water maze (Hashimoto et al. 2001), $TR\beta$ knockout mice performed equally well as wild-type controls in the water maze and no obvious neuroanatomical alterations were found in their hippocampus (Forrest et al. 1996; O'Shea and Williams 2002). This suggests that $TR\beta$ might exert very specific roles in hippocampal function, which are impaired by the presence of repressive apo-$TR\beta$ but not by the complete lack of $TR\beta$.

Hippocampal thyroid hormone signaling has also been implicated in epilepsy. While maternal hypothyroidism in humans and lack of $TR\beta$ in rodents have both been associated with increased susceptibility to a certain type of seizures (Andersen et al. 2013; Ng et al. 2001), mice expressing a mutant $TR\alpha1$ seem to be less susceptible to chemically induced seizures (Hadjab-Lallemend et al. 2010). Although the exact mechanisms are not being understood and might be different for the various types of seizures, these findings underline an important role for TRs in the hippocampal circuitry implicated in epilepsy.

7.4.3 Hypothalamus

The hypothalamus is located at ventral part of the forebrain, and considered the master regulator of homeostasis, controlling food intake, energy expenditure, and autonomic function. The most studied aspect of thyroid hormone action in the hypothalamus is the feedback regulation of the HPT-axis (hypothalamus–pituitary–thyroid axis). This function is governed by the thyrotropin-releasing hormone (TRH) containing hypophysiotropic neurons in the paraventricular nucleus (PVN), which control the activity of the thyroid-stimulating hormone (TSH) releasing cells of the pituitary (Zoeller et al. 2007). TRH is also expressed in other hypothalamic nuclei; however, only in the PVN is it regulated by thyroid hormone. This negative regulation of TRH expression is part of the HPT axis' negative feedback loop, and mediated by TRβ2 as demonstrated in knockout studies (O'Shea and Williams 2002). In addition to this acute regulation, thyroid hormone is also important for establishing the central setpoint of the HPT axis during development: inappropriately high levels of thyroxine during this critical period result in altered HPT axis function and abnormal regulation in mice and humans (Azizi et al. 1974; Bakke et al. 1974; Hernandez et al. 2006; Higuchi et al. 2005).

Another presumed role for hypothalamic TRβ signaling governs sexual behavior. Female mice lacking TRβ display increased sexual behavior, accompanied by an increase of oxytocin neurons in the PVN and a decrease in the preoptic area (Dellovade et al. 2000). In contrast, the number of oxytocin neurons was not affected in mice heterozygous for a mutant TRα1 (Mittag, unpublished).

More recently, it was demonstrated that a novel population of thermosensitive parvalbumin neurons in the anterior hypothalamic area also depends on intact TRα and TRβ signaling for proper development (Mittag et al. 2013). As these neurons regulate heart rate and blood pressure, they constitute a cellular link connecting developmental hypothyroidism to cardiovascular defects in the adult organism. In addition to controlling the autonomic output to the heart, hypothalamic thyroid hormone levels also regulate food intake and brown fat energy expenditure (Lopez et al. 2010; Sjogren et al. 2007). These actions of thyroid hormone rely on mTOR (mammalian target of rapamycin) and AMPK (AMP activated protein kinase) activation, respectively (Lopez et al. 2010; Varela et al. 2012); however, the exact pathways and target genes remain to be identified.

7.4.4 Cerebellum

The cerebellum develops as part of the hindbrain, and mainly controls motor function. As it is in mice highly dependent on thyroid hormone during postnatal development, the cerebellum is a perfect model system to dissect the hormone's action on a molecular and cellular level. Hypothyroidism in the cerebellum negatively affects the arborization of Purkinje neurons (Heuer and Mason 2003), GABAergic synaptogenesis (Fauquier et al. 2014) as well as the radial migration of the small granule neurons, with the consequence that the external granular layer persists longer. These neuronal defects are almost exclusively mediated by TRα1 (Fauquier et al. 2014; Heuer and Mason 2003), and consequently TRα1 mutant mice faithfully recapitulate all cerebellar defects found in congenital hypothyroidism (Fauquier et al. 2014; Flamant and Quignodon 2010; Wallis et al. 2008). That an unliganded TRα1 underlies these abnormalities is further confirmed by the observation that the negative consequences of hypothyroidism can be largely prevented by the deletion of TRα1, thereby removing TRα1 aporeceptor activity (Morte et al. 2002). Likewise, the detrimental damage of hyperthyroidism for cerebellar development found in an animal model lacking the thyroid hormone inactivating enzyme deiodinase type III can also be averted by TRα1 inactivation (Peeters et al. 2013). It is therefore well established that the aporeceptor activity of TRα1 is the main culprit for the defects found in the hypothyroid cerebellum postnatally; however, most cellular abnormalities normalize to a large extent later in life (Wallis et al. 2008), suggesting that their development is delayed but not completely abolished.

Not only neurons are affected by the lack of thyroid hormone in the cerebellum, Bergman glia cells also display abnormal morphology and mislocalized cell bodies in animal models replicating hypothyroidism (Fauquier et al. 2011). Moreover, early TRα1 action in Purkinje cells seems to be important for the secretion of specific neurotrophic factors that are required for oligodendrocyte precursor cell differentiation (Picou et al. 2012). Interestingly, in contrast to its permanent expression in postmitotic stellate and basket cells, TRα1 is only found in the early Purkinje neurons at P7 (Wallis et al. 2010). Later during maturation, these cells switch to TRβ (Bradley et al. 1992; Flamant and Gauthier 2013; Mellstrom et al. 1991; Wallis et al. 2010). Consequently, both TRα and TRβ mutant mice exhibit motoric problems as assessed by the rotarod—a test system aiming to quantify cerebellar function by analyzing the ability of mice to stay on a rotating rod (Hashimoto et al. 2001; Venero et al. 2005).

7.5 Brain Defects in Human Patients with a Mutant Thyroid Hormone Receptor

Given the important yet distinct functions of TRs in the developing and adult brain, specific psychomotor problems would be expected in humans with mutations in TRs. Mutations in TRβ were initially identified several decades ago; currently over 100 different mutations in more than 500 affected individuals are known (Cheng et al. 2010). And indeed, these patients display a resistance to thyroid hormone, with the most prominent symptom being high levels of circulating thyroid hormone due to the defective feedback in the HPT axis. This central hyperthyroidism complicates the interpretation of their phenotype, as specific aspects can be caused either by the defective TRβ or by the resulting activation of TRα1. Nevertheless, mutations in TRβ have been associated with hyperactivity disorders in 73 % of affected children (Brucker-Davis et al. 1995)—a finding supported by the hyperactive phenotype of transgenic mice carrying a mutation in TRβ (Wong et al. 1997). Moreover, several TRβ patients display a lower-than-average IQ as well as dyslexia (Cheng et al. 2010).

Mutations in TRα1 in humans have only been described recently, and to date three mutations in four different patients have been published (Bochukova et al. 2012; Moran et al. 2013; van Mullem et al. 2012, 2013). These patients display some features also observed in untreated developmental hypothyroidism, such as impaired motor coordination, gait, clumsiness, placid speech, and impaired cognition (Bochukova et al. 2012; Moran et al. 2013; van Mullem et al. 2012, 2013); however, their symptoms less pronounced than in patients with untreated congenital hypothyroidism. Moreover, similar to what was expected from the animal models (Vennstrom et al. 2008), childhood seizures have been reported in one patient, which persisted as chronic epilepsy into adulthood (Moran et al. 2013).

7.6 Conclusions

Based on the findings in humans and rodents, it is well established that TRs are of fundamental importance for brain development and function. Due to the limited overlap in their temporospatial expression patterns, TRα1 and TRβ exert very distinct functions in the brain with little if any overlap. TRα1 is likely more relevant for the brain than TRβ, given its widespread presence in almost all neurons; however, specific defects can also be assigned to impaired TRβ signaling especially in the hypothalamus.

In general, the interpretation of findings from mouse models with altered TRα1 or TRβ signaling is complicated by the fact that the adult animal displays a combination of acute and developmental defects. Moreover, although distinct tests such as the Morris water maze or the rotarod have been used, it is often not trivial to assign a molecular or anatomical defect to a defined and quantifiable phenotypic trait. This is especially true for studies in the field of thyroid hormone signaling, as several compensatory layers all working in concert, such as deiodinases, transporters, and feedback loops, exist to minimize the consequences of disturbances in the system. However, the constant refinement of available animal models, including conditional knockout and knockin strains or mice expressing tagged TR isoforms (Dudazy-Gralla et al. 2013; Flamant and Quignodon 2010; Wallis et al. 2010), will certainly contribute to a better understanding of the underlying processes at an increased molecular and cellular resolution. All available studies, however, have clearly demonstrated that an optimal level of maternal and fetal thyroid hormone is absolutely required to achieve optimal brain development. Therefore all measures should be taken to ensure proper thyroid hormone levels during pregnancy and later in life; especially since a sufficient supply with iodine is not even guaranteed in first world countries (Laurberg 2009; Vanderpump et al. 2011).

References

Andersen SL, Laurberg P, Wu CS et al (2013) Maternal thyroid dysfunction and risk of seizure in the child: a Danish nationwide cohort study. J Pregnancy 2013:636705

Anthony A, Adams PM, Stein SA (1993) The effects of congenital hypothyroidism using the hyt/hyt mouse on locomotor activity and learned behavior. Horm Behav 27(3):418–433

Azizi F, Vagenakis AG, Bollinger J et al (1974) Persistent abnormalities in pituitary function following neonatal thyrotoxicosis in the rat. Endocrinology 94(6):1681–1688

Bakke JL, Lawrence N, Wilber JF (1974) The late effects of neonatal hyperthyroidism upon the hypothalamic-pituitary-thyroid axis in the rat. Endocrinology 95(2):406–411

Beamer WJ, Eicher EM, Maltais LJ et al (1981) Inherited primary hypothyroidism in mice. Science 212(4490):61–63

Berbel P, Marco P, Cerezo JR et al (1996) Distribution of parvalbumin immunoreactivity in the neocortex of hypothyroid adult rats. Neurosci Lett 204(1–2):65–68

Berbel P, Mestre JL, Santamaria A et al (2009) Delayed neurobehavioral development in children born to pregnant women with mild hypothyroxinemia during the first month of gestation: the importance of early iodine supplementation. Thyroid 19(5):511–519

Berbel P, Navarro D, Auso E et al (2010) Role of late maternal thyroid hormones in cerebral cortex development: an experimental model for human prematurity. Cereb Cortex 20(6):1462–1475

Bernal J (2005) Thyroid hormones and brain development. Vitam Horm 71:95–122

Bernal J, Morte B (2013) Thyroid hormone receptor activity in the absence of ligand: physiological and developmental implications. Biochim Biophys Acta 1830(7):3893–3899

Bernal J, Pekonen F (1984) Ontogenesis of the nuclear 3,5,3'-triiodothyronine receptor in the human fetal brain. Endocrinology 114(2):677–679

Bochukova E, Schoenmakers N, Agostini M et al (2012) A mutation in the thyroid hormone receptor alpha gene. N Engl J Med 366(3):243–249

Bradley DJ, Young WS III, Weinberger C (1989) Differential expression of alpha and beta thyroid hormone receptor genes in rat brain and pituitary. Proc Natl Acad Sci U S A 86(18):7250–7254

Bradley DJ, Towle HC, Young WS III (1992) Spatial and temporal expression of alpha- and beta-thyroid hormone receptor mRNAs, including the beta 2-subtype, in the developing mammalian nervous system. J Neurosci 12(6):2288–2302

Brucker-Davis F, Skarulis MC, Grace MB et al (1995) Genetic and clinical features of 42 kindreds with resistance to thyroid hormone. The National Institutes of Health Prospective Study. Ann Intern Med 123(8):572–583

Carre JL, Demerens C, Rodriguez-Pena A et al (1998) Thyroid hormone receptor isoforms are sequentially expressed in oligodendrocyte lineage cells during rat cerebral development. J Neurosci Res 54(5):584–594

Chatonnet F, Picou F, Fauquier T et al (2011) Thyroid hormone action in cerebellum and cerebral cortex development. J Thyroid Res 2011:145762

Chatonnet F, Guyot R, Benoit G et al (2013) Genome-wide analysis of thyroid hormone receptors shared and specific functions in neural cells. Proc Natl Acad Sci U S A 110(8):E766–E775

Chen C, Zhou Z, Zhong M et al (2011) Excess thyroid hormone inhibits embryonic neural stem/progenitor cells proliferation and maintenance through STAT3 signalling pathway. Neurotox Res 20(1):15–25

Cheng SY, Leonard JL, Davis PJ (2010) Molecular aspects of thyroid hormone actions. Endocr Rev 31(2):139–170

de Escobar GM, Obregon MJ, del Rey FE (2007) Iodine deficiency and brain development in the first half of pregnancy. Public Health Nutr 10(12A):1554–1570

De Felice M, Di Lauro R (2004) Thyroid development and its disorders: genetics and molecular mechanisms. Endocr Rev 25(5):722–746

de Guglielmone AE, Gomez CJ (1966) Free amino acids in different areas of rat brain. Acta Physiol Lat Am 16(1):26–31

Dellovade TL, Chan J, Vennstrom B et al (2000) The two thyroid hormone receptor genes have opposite effects on estrogen-stimulated sex behaviors. Nat Neurosci 3(5):472–475

DeLong GR, Stanbury JB, Fierro-Benitez R (1985) Neurological signs in congenital iodine-deficiency disorder (endemic cretinism). Dev Med Child Neurol 27(3):317–324

Dudazy-Gralla S, Nordstrom K, Hofmann PJ et al (2013) Identification of thyroid hormone response elements in vivo using mice expressing a tagged thyroid hormone receptor alpha 1. Biosci Rep 33(2):e00027

Fauquier T, Romero E, Picou F et al (2011) Severe impairment of cerebellum development in mice expressing a dominant-negative mutation inactivating thyroid hormone receptor alpha1 isoform. Dev Biol 356(2):350–358

Fauquier T, Chatonnet F, Picou F et al (2014) Purkinje cells and Bergmann glia are primary targets of the TRalpha1 thyroid hormone receptor during mouse cerebellum postnatal development. Development 141(1):166–175

Flamant F, Gauthier K (2013) Thyroid hormone receptors: the challenge of elucidating isotype-specific functions and cell-specific response. Biochim Biophys Acta 1830(7):3900–3907

Flamant F, Quignodon L (2010) Use of a new model of transgenic mice to clarify the respective functions of thyroid hormone receptors in vivo. Heart Fail Rev 15(2):117–120

Flamant F, Poguet AL, Plateroti M et al (2002) Congenital hypothyroid Pax8(-/-) mutant mice can be rescued by inactivating the TRalpha gene. Mol Endocrinol 16(1):24–32

Forrest D, Erway LC, Ng L et al (1996) Thyroid hormone receptor beta is essential for development of auditory function. Nat Genet 13(3):354–357

Ghassabian A, Bongers-Schokking JJ, Henrichs J et al (2011) Maternal thyroid function during pregnancy and behavioral problems in the offspring: the generation R study. Pediatr Res 69(5 Pt 1):454–459

Gilbert ME, Lasley SM (2013) Developmental thyroid hormone insufficiency and brain development: a role for brain-derived neurotrophic factor (BDNF)? Neuroscience 239:253–270

Gilbert ME, Sui L, Walker MJ et al (2007) Thyroid hormone insufficiency during brain development reduces parvalbumin immunoreactivity and inhibitory function in the hippocampus. Endocrinology 148(1):92–102

Gil-Ibanez P, Morte B, Bernal J (2013) Role of thyroid hormone receptor subtypes alpha and beta on gene expression in the cerebral cortex and striatum of postnatal mice. Endocrinology 154(5):1940–1947

Goodman JH, Gilbert ME (2007) Modest thyroid hormone insufficiency during development induces a cellular malformation in the corpus callosum: a model of cortical dysplasia. Endocrinology 148(6):2593–2597

Gothe S, Wang Z, Ng L et al (1999) Mice devoid of all known thyroid hormone receptors are viable but exhibit disorders of the pituitary-thyroid axis, growth, and bone maturation. Genes Dev 13(10):1329–1341

Guadano-Ferraz A, Benavides-Piccione R, Venero C et al (2003) Lack of thyroid hormone receptor alpha1 is associated with selective alterations in behavior and hippocampal circuits. Mol Psychiatry 8(1):30–38

Hadjab-Lallemend S, Wallis K, van Hogerlinden M et al (2010) A mutant thyroid hormone receptor alpha1 alters hippocampal circuitry and reduces seizure susceptibility in mice. Neuropharmacology 58(7):1130–1139

Hashimoto K, Curty FH, Borges PP et al (2001) An unliganded thyroid hormone receptor causes severe neurological dysfunction. Proc Natl Acad Sci U S A 98(7):3998–4003

Hernandez A, Martinez ME, Fiering S et al (2006) Type 3 deiodinase is critical for the maturation and function of the thyroid axis. J Clin Invest 116(2):476–484

Heuer H, Mason CA (2003) Thyroid hormone induces cerebellar Purkinje cell dendritic development via the thyroid hormone receptor alpha1. J Neurosci 23(33):10604–10612

Higuchi R, Miyawaki M, Kumagai T et al (2005) Central hypothyroidism in infants who were born to mothers with thyrotoxicosis before 32 weeks' gestation: 3 cases. Pediatrics 115(5):e623–e625

Horn S, Heuer H (2010) Thyroid hormone action during brain development: more questions than answers. Mol Cell Endocrinol 315(1–2):19–26

Iskaros J, Pickard M, Evans I et al (2000) Thyroid hormone receptor gene expression in first trimester human fetal brain. J Clin Endocrinol Metab 85(7):2620–2623

Itoh Y, Esaki T, Kaneshige M et al (2001) Brain glucose utilization in mice with a targeted mutation in the thyroid hormone alpha or beta receptor gene. Proc Natl Acad Sci U S A 98(17):9913–9918

Laurberg P (2009) Thyroid function: thyroid hormones, iodine and the brain-an important concern. Nat Rev Endocrinol 5(9):475–476

Liu D, Teng W, Shan Z et al (2010) The effect of maternal subclinical hypothyroidism during pregnancy on brain development in rat offspring. Thyroid 20(8):909–915

Lopez M, Varela L, Vazquez MJ et al (2010) Hypothalamic AMPK and fatty acid metabolism mediate thyroid regulation of energy balance. Nat Med 16(9):1001–1008

Mai W, Janier MF, Allioli N et al (2004) Thyroid hormone receptor alpha is a molecular switch of cardiac function between fetal and postnatal life. Proc Natl Acad Sci U S A 101(28):10332–10337

Mansouri A, Chowdhury K, Gruss P (1998) Follicular cells of the thyroid gland require Pax8 gene function. Nat Genet 19(1):87–90

Mellstrom B, Naranjo JR, Santos A et al (1991) Independent expression of the alpha and beta c-erbA genes in developing rat brain. Mol Endocrinol 5(9):1339–1350

Mittag J, Friedrichsen S, Heuer H et al (2005) Athyroid Pax8-/- mice cannot be rescued by the inactivation of thyroid hormone receptor alpha1. Endocrinology 146(7):3179–3184

Mittag J, Lyons DJ, Sallstrom J et al (2013) Thyroid hormone is required for hypothalamic neurons regulating cardiovascular functions. J Clin Invest 123(1):509–516

Moran C, Schoenmakers N, Agostini M et al (2013) An adult female with resistance to thyroid hormone mediated by defective thyroid hormone receptor alpha. J Clin Endocrinol Metab 98(11):4254–4261

Morte B, Manzano J, Scanlan T et al (2002) Deletion of the thyroid hormone receptor alpha 1 prevents the structural alterations of the cerebellum induced by hypothyroidism. Proc Natl Acad Sci U S A 99(6):3985–3989

Ng L, Pedraza PE, Faris JS et al (2001) Audiogenic seizure susceptibility in thyroid hormone receptor beta-deficient mice. Neuroreport 12(11):2359–2362

O'Shea PJ, Williams GR (2002) Insight into the physiological actions of thyroid hormone receptors from genetically modified mice. J Endocrinol 175(3):553–570

Oppenheimer JH, Schwartz HL, Lane JT et al (1991) Functional relationship of thyroid hormone-induced lipogenesis, lipolysis, and thermogenesis in the rat. J Clin Invest 87(1):125–132

Pathak A, Sinha RA, Mohan V et al (2011) Maternal thyroid hormone before the onset of fetal thyroid function regulates reelin and downstream signaling cascade affecting neocortical neuronal migration. Cereb Cortex 21(1):11–21

Peeters RP, Hernandez A, Ng L et al (2013) Cerebellar abnormalities in mice lacking type 3 deiodinase and partial reversal of phenotype by deletion of thyroid hormone receptor alpha1. Endocrinology 154(1):550–561

Perez-Castillo A, Bernal J, Ferreiro B et al (1985) The early ontogenesis of thyroid hormone receptor in the rat fetus. Endocrinology 117(6):2457–2461

Picou F, Fauquier T, Chatonnet F et al (2012) A bimodal influence of thyroid hormone on cerebellum oligodendrocyte differentiation. Mol Endocrinol 26(4):608–618

Quignodon L, Legrand C, Allioli N et al (2004) Thyroid hormone signaling is highly heterogeneous during pre- and postnatal brain development. J Mol Endocrinol 33(2):467–476

Rodriguez-Pena A (1999) Oligodendrocyte development and thyroid hormone. J Neurobiol 40(4):497–512

Salto C, Kindblom JM, Johansson C et al (2001) Ablation of TRalpha2 and a concomitant overexpression of alpha1 yields a mixed hypo- and hyperthyroid phenotype in mice. Mol Endocrinol 15(12):2115–2128

Schwartz HL, Strait KA, Ling NC et al (1992) Quantitation of rat tissue thyroid hormone binding receptor isoforms by immunoprecipitation of nuclear triiodothyronine binding capacity. J Biol Chem 267(17):11794–11799

Sharlin DS, Gilbert ME, Taylor MA et al (2010) The nature of the compensatory response to low thyroid hormone in the developing brain. J Neuroendocrinol 22(3):153–165

Sjogren M, Alkemade A, Mittag J et al (2007) Hypermetabolism in mice caused by the central action of an unliganded thyroid hormone receptor alpha1. EMBO J 26(21):4535–4545

Tinnikov A, Nordstrom K, Thoren P et al (2002) Retardation of post-natal development caused by a negatively acting thyroid hormone receptor alpha1. EMBO J 21(19):5079–5087

van Mullem A, van Heerebeek R, Chrysis D et al (2012) Clinical phenotype and mutant TRalpha1. N Engl J Med 366(15):1451–1453

van Mullem AA, Chrysis D, Eythimiadou A et al (2013) Clinical phenotype of a new type of thyroid hormone resistance caused by a mutation of the TRalpha1 receptor; consequences of LT4 treatment. J Clin Endocrinol Metab 98(7):3029–3038

Vanderpump MP, Lazarus JH, Smyth PP et al (2011) Iodine status of UK schoolgirls: a cross-sectional survey. Lancet 377(9782):2007–2012

Varela L, Martinez-Sanchez N, Gallego R et al (2012) Hypothalamic mTOR pathway mediates thyroid hormone-induced hyperphagia in hyperthyroidism. J Pathol 227(2):209–222

Venero C, Guadano-Ferraz A, Herrero AI et al (2005) Anxiety, memory impairment, and locomotor dysfunction caused by a mutant thyroid hormone receptor alpha1 can be ameliorated by T3 treatment. Genes Dev 19(18):2152–2163

Vennstrom B, Mittag J, Wallis K (2008) Severe psychomotor and metabolic damages caused by a mutant thyroid hormone receptor alpha 1 in mice: can patients with a similar mutation be found and treated? Acta Paediatr 97(12):1605–1610

Wallis K, Sjogren M, van Hogerlinden M et al (2008) Locomotor deficiencies and aberrant development of subtype-specific GABAergic interneurons caused by an unliganded thyroid hormone receptor alpha1. J Neurosci 28(8):1904–1915

Wallis K, Dudazy S, van Hogerlinden M et al (2010) The thyroid hormone receptor alpha1 protein is expressed in embryonic postmitotic neurons and persists in most adult neurons. Mol Endocrinol 24(10):1904–1916

Wang S, Teng W, Gao Y et al (2012) Early levothyroxine treatment on maternal subclinical hypothyroidism improves spatial learning of offspring in rats. J Neuroendocrinol 24(5):841–848

Wiens SC, Trudeau VL (2006) Thyroid hormone and gamma-aminobutyric acid (GABA) interactions in neuroendocrine systems. Comp Biochem Physiol A Mol Integr Physiol 144(3):332–344

Wikstrom L, Johansson C, Salto C et al (1998) Abnormal heart rate and body temperature in mice lacking thyroid hormone receptor alpha 1. EMBO J 17(2):455–461

Wong R, Vasilyev VV, Ting YT et al (1997) Transgenic mice bearing a human mutant thyroid hormone beta 1 receptor manifest thyroid function anomalies, weight reduction, and hyperactivity. Mol Med 3(5):303–314

Zhang L, Cooper-Kuhn CM, Nannmark U et al (2010) Stimulatory effects of thyroid hormone on brain angiogenesis in vivo and in vitro. J Cereb Blood Flow Metab 30(2):323–335

Zoeller RT, Tan SW, Tyl RW (2007) General background on the hypothalamic-pituitary-thyroid (HPT) axis. Crit Rev Toxicol 37(1–2):11–53

Chapter 8
Using Mouse Genetics to Investigate Thyroid Hormone Signaling in the Developing and Adult Brain

F. Chatonnet, S. Richard, and F. Flamant

Abstract The developing and adult brain is a main target organ of thyroid hormones (including the prohormone thyroxine and its active derivative tri-iodothyronine or T3). Mouse genetics offers a number of promising possibilities to study their pleiotropic influence on the central nervous system, and to distinguish it from their peripheral function. In the following, we review recent advances brought by mouse genetics in our understanding of thyroid hormone signaling in the brain, both during development and in the adult. We particularly emphasize on the latest findings about thyroid hormone transporters and synthesis pathway which bring a new view on the regulation of thyroid hormone levels sensed by brain cells. Roles of the thyroid hormone receptors, which have been reviewed elsewhere are only briefly discussed.

Keywords Brain development • Transporters • Receptors • Deiodinases

8.1 T3 Synthesis and Transport

All neural cells seem to possess at least one of the TRα1, TRβ1, or TRβ2 thyroid hormone nuclear receptors, encoded in mice by the two *Thra* and *Thrb* genes, and thus have the ability to respond to T3 stimulation throughout development and adulthood (Williams 2008). As the receptors act as ligand-gated transcription factors, the cellular response primarily consists in the transcriptional regulation of a number of genes, usually called "target genes." However T3 access to these receptors is highly regulated. In the course of development, thyroid hormone receptors

F. Chatonnet • S. Richard • F. Flamant (✉)
Institut de Génomique Fonctionnelle de Lyon, Université de Lyon, Université Lyon 1, INRA, CNRS, Ecole Normale Supérieure de Lyon, 46 allée d'Italie, 69364 Lyon Cedex 07, France
e-mail: fabrice.chatonnet@ens-lyon.fr; sabine.richard@ens-lyon.fr;
frederic.flamant@ens-lyon.fr

© Springer Science+Business Media New York 2016
N. Koibuchi, P.M. Yen (eds.), *Thyroid Hormone Disruption and Neurodevelopment*,
Contemporary Clinical Neuroscience, DOI 10.1007/978-1-4939-3737-0_8

expression can be detected much earlier than T3. For instance, *Thra* is expressed before neurulation (Chassande et al. 1997), whereas T3 is found in the fetal brain from mid-gestation, around embryonic day 15 (E15) in mice. This suggests that TRα target genes are repressed at earlier stages, since unliganded TRα1 is known to inhibit gene expression (Flamant et al. 2002; Morte et al. 2002). In fact, T3 is detected in the fetal brain 2 days before the onset of fetal thyroid gland function, indicating that the onset of T3 signaling in brain results from local deiodination of maternal thyroxine by type 2 deiodinase (D2 encoded by the *Dio2* gene), thyroxine being more easily transferred through the brain–blood barrier than maternal T3 (Morte et al. 2002; Quignodon et al. 2004); Some brain cell types also express type 3 deiodinase (D3 encoded by the *Dio3* gene), an enzyme able to catabolize both T4 and T3. D2 is upregulated in brain in case of T3 deficiency, while D3 is activated when an excess of T3 is present (Barca-Mayo et al. 2011). Therefore the brain possesses an additional feedback regulation, which makes it more resistant than other organs to variations in T3 serum levels.

The transfer of T3 to neurons and glial cells is also dependent on a number of transporters, which display different expression patterns (Kinne et al. 2010; Schweizer and Kohrle 2013). These include the monocarboxylate transporter 8 (MCT8, alias SLC16A2), the neuron-specific (Grijota-Martinez et al. 2011) organic anion transporter 14 (OATP14/OATP1C1/SLCO1C1), and L-type amino acid transporter 1 and 2 (LAT1/SLC7A5 and LAT2/SLC7A6). Furthermore, the nicotinamide adenine dinucleotide phosphate-dependent cytosolic T3-binding protein CRYM/µ-crystallin is required to prevent rapid efflux of T3 from the brain (Suzuki et al. 2007; Lang et al. 2011). Transthyretin, a thyroxine-binding protein of the serum, is also present in the cerebrospinal fluid and in the choroid plexus, which is suggestive of its possible involvement in the brain distribution of thyroxine. However, detailed analysis of knockout mice does not provide any support to this hypothesis (Palha et al. 2000; Sousa et al. 2005). Local metabolism and differential transport have thus the potential to generate a very heterogeneous pattern of T3 distribution (Pinna et al. 2002) and T3 signaling (Chassande et al. 1997) which is not necessarily correlated with serum content. In order to identify the components of this system with predominant function, several genetically modified mouse strains have been generated (Table 8.1).

Surprisingly the effect of *Dio2* gene knockout on brain T3 content is moderate and temporary, a 50 % reduction being found only in juveniles cerebellum (Liao et al. 2011; Galton et al. 2007, 2009). This may seem sufficient to entail visible defects at the molecular or cellular level, but this is not the case, suggesting that a modified T3 distribution limits the adverse effects of D2 deficiency. *Dio3* gene knockout only results in a progressive accumulation of T3 in the anterior cortex (Hernandez et al. 2010; Barca-Mayo et al. 2011).

The case of MCT8 is puzzling: mutations of the human gene cause the Allen–Herndon–Dudley syndrome, associating a very severe mental retardation resembling neurological cretinism to apparent hyperthyroidism in peripheral organs. This suggests that MCT8 function is crucial for T3 transfer in the fetal brain (Heuer and Visser 2013). However *Mct8* gene knockout mice have a very modest brain

Table 8.1 Mouse mutant strains used for study of thyroid hormone role in CNS development and function

Strain	Type of mutation	Protein/function affected	Effect on T3 signaling	CNS phenotype
$Dio2^{-/-}$	Knockout	Deiodinase type 2, T3 production	Reduce T3 production	Mild CNS phenotype
$Dio3^{-/-}$	Knockout	Deiodinase type 3, T3 and T4 degradation	Reduce T3 degradation	Mild CNS phenotype
$Slco1c1^{-/-}$ (Oatp1c1)	Knockout	Transporter	Reduce T3 transport in neural cells	Mild CNS phenotype
$Slc7a5^{-/-}$ (Lat1)	Knockout	Transporter	Reduce T3 transport in neural cells	Mild CNS phenotype
$Slc7a6^{-/-}$ (Lat2)	Knockout	Transporter	Reduce T3 transport in neural cells	Mild CNS phenotype
$Mct8^{-/-}$	Knockout	Transporter	Reduce T3 transport in neural cells	Mild CNS phenotype
$Dio2^{-/-};Mct8^{-/-}$	Compound knockout	T3 production and transport	Reduce T3 transport and production in neural cells	Cx gene expression modified
$TR\alpha^{+/m}$	Knock-in	Receptor-α1	Low affinity for T3. Partial dominant-negative effect (overruled by excess T3)	Hypothalamic, cortical and cerebellar defects
$TR\alpha^{AMI}$	Knock-in, cre/lox dependent	Receptor-α1	Dominant-negative, no transcriptional activation	Hypothyroid cerebellum
$TR\alpha1^{GFP}$	Knock-in	Receptor-α1	Fusion protein with GFP	Detection of TRα1 expression
$TR\beta^{\Delta337T}$	Knock-in of a human RTH mutation	Receptor-β1	Low affinity for T3	Cerebellar defects different from hypothyroidism

phenotype. They do not display histological signs of hypothyroidism (Trajkovic et al. 2007) and adult behavior tests can only reveal subtle defects (Wirth et al. 2009). Introducing *Mct8* knockout allele in transgenic reporter mice (Quignodon et al. 2004) does not alter the early expression of the reporter construct in an obvious manner (unpublished observation). Although direct measurement indicates a signification reduction of T3 transport in *Mct8* knockout mice (Wirth et al. 2009), the brain T3 content is only moderately affected. A perinatal increase is followed by a permanent decrease (Ferrara et al. 2013). That a defect in T3 transport has limited effect on intra-cerebral content is explained by the predominant transport of the prohormone T4 to the brain, which is not affected by the mouse MCT8 mutation (Ceballos et al. 2009). One possibility would be that compensatory response of other transporters and deiodinases attenuate *Mct8* knockout mice phenotype (Liao et al. 2011; Ferrara et al. 2013). It has also been suggested that different consequences of MCT8 mutations in humans and mice reflect differences in the timing of transporter expression. Earlier expression of LAT2 (Braun et al. 2011a) or OATP14 (Mayerl et al. 2012) would provide a compensation in mice but not in humans.

Lat2 knockout does not alone alter T3 brain content (Braun et al. 2011b). The effect of *Oatp14* knockout on brain average T3 content is moderate, but this again can be explained compensatory deregulation of deiodinases. Interestingly, specific brain areas seem to be more sensitive to the mutation (Mayerl et al. 2012). Combining these mutations with *Mct8* knockout will certainly bring valuable information in the near future.

Taken together, these genetic data suggest that T3 can access to neurons by several routes and that gene functions are partially redundant. In line with this idea, combining *Dio2* and *Mct8* mutations have a synergistic effect on gene expression in cortex (Liao et al. 2011; Morte et al. 2010). This suggests the existence of two pathways for T3 entry in the cortex: one relies on MCT8-mediated transport in neurons; the other is MCT8-independent and involves T4 deiodination by D2 in astrocytes. T3 regulates gene expression in cortex both positively and negatively. The molecular mechanism sustaining the negative gene regulation by T3 is not known, and why the knockout of *Dio2*, but not of *Mct8*, alters it without changing the expression level of positively regulated genes remains unexplained (Hernandez et al. 2012).

In conclusion, the transport and metabolism of T3 in mouse fetal and postnatal brain seems to be a very robust system, with various possibilities for compensation. The human brain also expresses a wide array of transporters from the first gestation trimester (Chan et al. 2003; Alkemade et al. 2011). This outlines the paradoxical severity of the human Allen–Herndon–Dudley neurological syndrome, which is due to a single MCT8 mutation. A testable hypothesis would be that thyroxine transport is affected during early human pregnancy, but not during mouse gestation, because other thyroxine transporters may be expressed in the mouse brain at earlier developmental stages.

8.2 T3 Functions in Postnatal Cerebellum

Cerebellum development, which is mainly postnatal, has been the favored model to study the neurodevelopmental consequences of hypothyroidism (Koibuchi 2013). Other brain areas that develop earlier also need T3 for proper development (Howdeshell 2002), suggesting that specific steps of neuronal differentiation require T3 stimulation in all brain areas. However, as they were based on inconvenient animals models, only few attempts have been made to precisely study these events at the cellular and molecular level outside of the cerebellum (Morte et al. 2010; Navarro et al. 2014). Whether the underlying genetic program is the same in different brain areas remains thus unclear, although the repertoire of T3 responsive genes seems to differ widely (Morte et al. 2010; Gil-Ibanez et al. 2013; Navarro et al. 2014).

Perinatal T3 deficiency alter different differentiation processes in cerebellum neuronal cells: the most obvious histological sign is the persistence of the external granular layer, a transient structure where granule cell progenitors proliferate. The accumulation of granule cell progenitors in the layer reflects their slow proliferation, combined with impaired inward migration. The maturation of GABAergic interneurons, which populate the molecular layer, is also retarded. Purkinje neurons display reduced dendritic growth and synaptogenesis. T3 deficiency also affects glial cells, increasing astrocytes proliferation, retarding oligodendrocytes differentiation and myelin formation. The radial processes of Bergmann glia have an abnormal morphology. As all these cytological defects are accompanied by a reduction in the production of neurotrophin 3, nerve growth factor, insulin-like growth factor, Sonic hedgehog, brain-derived neurotrophic factor, and galanin, the possibility exists that a large fraction of these effects are not cell type-autonomous consequences of the presence of unliganded TRs, but secondary to the impaired production of growth factors. In vitro, primary cell cultures can be used to isolate each cerebellum cell type from its microenvironment and address its ability to display a direct response to T3 stimulation. This is the case for granule cells (Chatonnet et al. 2012), Purkinje neurons (Heuer and Mason 2003; Boukhtouche et al. 2010) astrocytes (Mendes-de-Aguiar et al. 2008), and oligodendrocyte precursor cells (OPCs (Durand and Raff 2000)).

In vivo, a TRα1 dominant-negative mutation, dependent on the Cre/loxP technology for its expression (Table 8.1), is sufficient to phenocopy congenital hypothyroidism in cerebellum (Fauquier et al. 2011). In the case of OPCs (Picou et al. 2012), CNS ubiquitous mutation delays OPCs differentiation and myelin formation. However, the restriction of the mutation to OPCs using the Cre/loxP system has no immediate effect, while expressing the mutation only in Purkinje cells or astrocytes delays OPCs differentiation. This is consistent with an indirect effect of T3 on myelin formation, relayed by Purkinje cells and astrocytes. However, expressing TRα1 mutation in OPCs has long-term effect. The white matter of adult mutant mice cerebellum is progressively invaded by slow-cycling OPCs, suggesting that the in vitro observations correspond to a distinct effect and direct effect on cell cycle taking place in the adult population of OPCs. This experimental strategy is thus promising to unravel the influence of T3 on the network of cellular interactions that govern normal development and maintenance of brain cell types.

The respective functions of TRα1 and TRβ1 in cerebellum are currently unclear. TRα1 is probably present in all cell types (Bradley et al. 1989) but tends to accumulate in postmitotic neurons, as has been shown with a TRα1-GFP fusion reporter mouse strain (Wallis et al. 2010). Nevertheless, cerebellar hypothyroidism is recapitulated by the ubiquitous expression of a TRα1 dominant-negative mutation, as mentioned previously, suggesting that TRα1 exerts a major role in cerebellar development. TRβ1 is present mainly in Purkinje neurons, but a point-mutation is sufficient to exert a global effect on cerebellum development (Portella et al. 2010). However the mutation also increases the circulating level of T3, by impairing hypothalamic and pituitary feedback-regulation, and it is likely that this indirectly accelerates cerebellum development. Whereas the Purkinje cells phenotype is probably a direct effect of the mutation, the observed foliation defect is possibly a consequence of a local excess in T3. This conclusion is based on similarity with *Dio3* gene knockout (Peeters et al. 2013) and hyperthyroidism (Lauder et al. 1974) phenotypes. It is reinforced by the fact that this defect can be reverted by TRα1 knockout (Peeters et al. 2013).

8.3 T3 Function in the Adult Hypothalamus

Hypothalamus is a place where T3 exerts feedback regulation on TRH, but T3 has been proposed to participate to many other integrative functions in this brain area: circadian rhythm (Herwig et al. 2009, 2013), seasonal perception (Yoshimura 2013), chronic inflammatory response (Boelen et al. 2011), appetite (Varela et al. 2012), energy metabolism (Coppola et al. 2007; Fliers et al. 2009; Lopez et al. 2010), autonomic control of body temperature, cardiac rhythm (Mittag et al. 2010), etc. Both D2 and D3 deiodinases are present in hypothalamus and their multiple modes of regulation bring the possibility for a very precise and rapid control of local T3 signaling. Mouse genetics will certainly bring major contribution is unraveling this complex system, although few attempts have been made to date, due to lack of cellular populations markers and dedicated mutant strains.

One of the most enigmatic cell-type in the hypothalamus is a class of specialized astrocytes known as tanycytes, which border the ependymal zone of the third ventricle and are therefore in direct contact with the cerebrospinal fluid, the hypothalamic blood circulation, and surrounding neurons/neuronal processes. These cells have been shown to express a number of thyroid hormone signaling components, such as D2 and MCT8, and it is suggested that they finely and rapidly regulate the local concentration in thyroid hormone, therefore fine-tuning thyroid hormone signaling intensity perceived by surrounding cells (Fonseca et al. 2013). They seem also able to directly modify TRH levels after its release by hypothalamic TRH neurons (Sanchez et al. 2009), providing another level of regulation in the already complex negative feedback-regulation of thyroid hormone secretion. Unfortunately, there is no specific genetic tool available to decipher the role of the tanycytes more precisely.

The respective function of TRα1, TRβ1, and TRβ2 is not yet clearly defined. TRβ2 is the predominant regulator of TRH feedback regulation (Abel et al. 2001). Surprisingly, in the same cells, both TRα1 and TRβ1/2 KO alter the negative regulation of MC4R by T3 (Decherf et al. 2010). This gene encodes type 4 melanocortin receptor, which is thought to link the autonomic control of energy metabolism to peripheral reserve status.

It has also been proposed that the cardiovascular and hypermetabolic phenotype of mice with a TRα1 mutation that severely reduces T3 affinity (Tinnikov et al. 2002) (TRα$^{+/m}$ Table 8.1) originates in the hypothalamus (Mittag et al. 2010), although alternative explanation can hardly be ruled out. Interestingly, expression of this TRα1 mutation eliminates a previously unknown category of parvalbumin-expressing neurons, with an apparent role in autonomic control of blood pressure and heart rate. The activity of these neurons is thermosensitive, suggesting a possible link between cardiovascular function, core temperature, and T3 hypothalamic function (Mittag et al. 2013).

In summary, T3 action and regulation in the hypothalamus has only begun to be investigated, but the lack of genetic mouse strains targeting identified cellular populations hindered the understanding of this complex system. In the future and given the central role of the hypothalamus in controlling diverse aspects of the metabolism and energy expenditure, which are linked to major diseases in western countries, it will be of crucial interest to develop such genetic tools.

8.4 Involvement of T3 Signaling in Mood and Cognitive Function in Adults

In humans, adult onset of either hyperthyroidism or hypothyroidism is associated with changes in emotional behavior and intellectual performance. Hypothyroid patients frequently exhibit alterations in concentration, memory, psychomotor speed, and increased rates of depressive and anxiety disorders (Heinrich and Grahm 2003; Sait Gonen et al. 2004; Bauer et al. 2008). In hyperthyroidism, anxiety disorders have been found to occur in approximately 60 % of the patients while depressive disorders have been reported in 31–69 % (Hage and Azar 2012). Moreover, variation of serum free-T4, even within the physiological range, correlates with cognitive performance (Beydoun et al. 2013). In older people in particular, there is a substantial body of evidence to support the association of subclinical hypothyroidism and cognitive impairment (Gan and Pearce 2012). However, there is no clear mechanistic explanation for these associations.

Paralleling the associations reported in humans, adult-onset hypothyroidism has been found to influence rodent cognitive and emotional behavior (Montero-Pedrazuela et al. 2003, 2006; Vallortigara et al. 2009). Genetic manipulation of thyroid hormone signaling also induces alterations in mouse cognitive and emotional behavior. Thus, impaired cognitive function has been reported in TRβ mutant mice: TRβ^{Δ337T} mice exhibited a learning deficit in the Morris water maze (Hashimoto et al. 2001), whereas TRβ^{PV} mice were characterized by hyperactivity, impulsivity, impaired attention, and impaired learning in a vigilance task (Siesser et al. 2006). Male TRβ knockout mice also exhibited signs of hyperactivity in the open-field (McDonald et al. 1998) and reduced anxiety in the elevated-plus maze and light-dark box tests (Vasudevan et al. 2013). By contrast, TRα1 knockout mice showed increased anxiety in the elevated-plus maze and open-field tests and reduced extinction in the contextual fear conditioning paradigm (Guadano-Ferraz et al. 2003; Vasudevan et al. 2013). In $TR\alpha^{0/0}$ mice, learning deficits were recorded in the Morris water maze but it was suggested that these deficits were related to increased anxiety, these mice exhibiting increased thigmotaxis in this task; $TR\alpha^{0/0}$ mice also exhibited increased anxiety in the open-field test (Wilcoxon et al. 2007). As expected considering the transcriptional effects of TRα1 in its aporeceptor state, $TR\alpha 1^{+/m}$ mice displayed a stronger behavioral phenotype than TRα knockout models, including extreme anxiety, as assessed in the open-field, the elevated-plus maze (Venero et al. 2005), and the light-dark box tests (Pilhatsch et al. 2010). These mice also exhibited reduced object recognition memory (Venero et al. 2005) and increased depressive-like behavior as evidenced in the shuttle box paradigm (Pilhatsch et al. 2010). The fact that treatment of adult $TR\alpha 1^{+/m}$ mice with a high dose of T3 alleviated all the aforementioned behavioral effects suggested that these effects were not of developmental origin and resulted from altered T3 signaling in the adult brain. Finally, *Mct8* knockout mice, which exhibit abnormally high plasma T3 levels but no obvious neurological defect, displayed decreased anxiety-related behavior in the modified hole-board test (Trajkovic et al. 2007; Wirth et al. 2009). This effect appears to be consistent with developmental effects of hyperthyroidism, since rats transiently treated with thyroxin after birth also exhibit reduced anxiety when tested as adults in the elevated-plus maze test (Yilmazer-Hanke et al. 2004).

More work is needed to unravel the mechanisms by which thyroid hormone signaling affects behavior. For mood and cognition as for other functions it appears that TRα1 and TRβ1 play different roles in the regulation of adult behavior, since a deficit in a particular receptor type cannot be compensated by the other type. It is thus likely that different mechanisms will be found for the two receptor types. TRα1 and TRβ1 are involved in the modulation of adult neural stem cell proliferation and differentiation (Lemkine et al. 2005; Montero-Pedrazuela et al. 2006; Kapoor et al. 2011; Lopez-Juarez et al. 2012). As a consequence, one of the possible ways of action of T3 on emotional and cognitive behavior involves modulation of hippocampal neurogenesis, which appears to be tightly related to depressive disorders and their treatment (Balu and Lucki 2009; Tanti and Belzung 2013). GABA-ergic interneurons might constitute another key part of the puzzle. Indeed, adult TRα1 knockout mice and $TR\alpha 1^{+/m}$ mice exhibited a reduced density of parvalbumin-positive terminals in the CA1 field of the hippocampus (Guadano-Ferraz et al. 2003; Venero et al. 2005). In $TR\alpha 1^{+/m}$ mice this effect was normalized by adult T3 treatment (Venero et al. 2005). Further investigation revealed an altered maturation and function of the GABA-ergic hippocampal network in these mice, with an increased density of calretinin-positive cells in the motor cortex and dorsal hippocampus (Hadjab-Lallemend et al. 2010; Wallis et al. 2010).

8.5 Conclusion

Data gathered in mouse show that mouse genetics can provide useful information about the roles of thyroid hormone receptors in the development of the brain and its adult function. These studies are also useful to better understand the phenotypes associated with the mutations of the human THRA and THRB genes. Mutations of THRB lead to a well-described syndrome known as "resistance to thyroid hormone" (Dumitrescu and Refetoff 2013) characterized by high circulating T3 and T4 levels and normal TSH level. This condition has been associated with mental retardation, hyperthyroidism being probably the cause. More recently, several THRA mutations have been reported (Bochukova et al. 2012; van Mullem et al. 2012, 2013; Moran et al. 2013) leading to normal-to-high T3 circulating levels. In these cases, the mutations are inferred to disturb the ligand-binding domain, and the cofactor-binding surface of the receptor is associated with various level of mental retardation. This observation suggests that TRα1 is the major mediator of the control of brain development by thyroid hormone.

Overall, mouse genetics has provided and will continue to provide a great wealth of data to better understand the complex role of thyroid hormone and its receptors in brain development and function.

Acknowledgements We thank Beatriz Morte for critical reading of the manuscript. Work in our laboratory is supported by Agence Nationale de la Recherche (Thyrogenomic2).

References

Abel ED, Ahima RS et al (2001) Critical role for thyroid hormone receptor beta2 in the regulation of paraventricular thyrotropin-releasing hormone neurons. J Clin Invest 107(8):1017–1023

Alkemade A, Friesema EC et al (2011) Expression of thyroid hormone transporters in the human hypothalamus. J Clin Endocrinol Metab 96(6):E967–E971

Balu DT, Lucki I (2009) Adult hippocampal neurogenesis: regulation, functional implications, and contribution to disease pathology. Neurosci Biobehav Rev 33(3):232–252

Barca-Mayo O, Liao XH et al (2011) Thyroid hormone receptor alpha and regulation of type 3 deiodinase. Mol Endocrinol 25(4):575–583

Bauer M, Goetz T et al (2008) The thyroid-brain interaction in thyroid disorders and mood disorders. J Neuroendocrinol 20(10):1101–1114

Beydoun MA, Beydoun HA et al (2013) Thyroid hormones are associated with cognitive function: moderation by sex, race and depressive symptoms. J Clin Endocrinol Metab 98(8):3470–3481

Bochukova E, Schoenmakers N et al (2012) A mutation in the thyroid hormone receptor alpha gene. N Engl J Med 366(3):243–249

Boelen A, Kwakkel J et al (2011) Beyond low plasma T3: local thyroid hormone metabolism during inflammation and infection. Endocr Rev 32(5):670–693

Boukhtouche F, Brugg B et al (2010) Induction of early Purkinje cell dendritic differentiation by thyroid hormone requires RORalpha. Neural Dev 5:18

Bradley DJ, Young WS III et al (1989) Differential expression of alpha and beta thyroid hormone receptor genes in rat brain and pituitary. Proc Natl Acad Sci U S A 86(18):7250–7254

Braun D, Kinne A et al (2011a) Developmental and cell type-specific expression of thyroid hormone transporters in the mouse brain and in primary brain cells. Glia 59(3):463–471

Braun D, Wirth EK et al (2011b) Aminoaciduria, but normal thyroid hormone levels and signalling, in mice lacking the amino acid and thyroid hormone transporter Slc7a8. Biochem J 439(2):249–255

Ceballos A, Belinchon MM et al (2009) Importance of monocarboxylate transporter 8 for the blood-brain barrier-dependent availability of 3,5,3′-triiodo-L-thyronine. Endocrinology 150(5):2491–2496

Chan S, McCabe CJ et al (2003) Thyroid hormone responsiveness in N-Tera-2 cells. J Endocrinol 178(1):159–167

Chassande O, Fraichard A et al (1997) Identification of transcripts initiated from an internal promoter in the c-erbA alpha locus that encode inhibitors of retinoic acid receptor-alpha and triiodothyronine receptor activities. Mol Endocrinol 11(9):1278–1290

Chatonnet F, Guyot R et al (2012) Genome-wide search reveals the existence of a limited number of thyroid hormone receptor alpha target genes in cerebellar neurons. PLoS One 7(5):e30703

Coppola A, Liu ZW et al (2007) A central thermogenic-like mechanism in feeding regulation: an interplay between arcuate nucleus T3 and UCP2. Cell Metab 5(1):21–33

Decherf S, Seugnet I et al (2010) Thyroid hormone exerts negative feedback on hypothalamic type 4 melanocortin receptor expression. Proc Natl Acad Sci U S A 107(9):4471–4476

Dumitrescu AM, Refetoff S (2013) The syndromes of reduced sensitivity to thyroid hormone. Biochim Biophys Acta 1830(7):3987–4003

Durand B, Raff M (2000) A cell-intrinsic timer that operates during oligodendrocyte development. Bioessays 22(1):64–71

Fauquier T, Romero E et al (2011) Severe impairment of cerebellum development in mice expressing a dominant-negative mutation inactivating thyroid hormone receptor alpha1 isoform. Dev Biol 356(2):350–358

Ferrara AM, Liao XH et al (2013) Changes in thyroid status during perinatal development of MCT8-deficient male mice. Endocrinology 154(7):2533–2541

Flamant F, Poguet AL et al (2002) Congenital hypothyroid Pax8(-/-) mutant mice can be rescued by inactivating the TRalpha gene. Mol Endocrinol 16(1):24–32

Fliers E, Klieverik LP et al (2009) Novel neural pathways for metabolic effects of thyroid hormone. Trends Endocrinol Metab 21(4):230–236

Fonseca TL, Correa-Medina M et al (2013) Coordination of hypothalamic and pituitary T3 production regulates TSH expression. J Clin Invest 123(4):1492–1500

Galton VA, Wood ET et al (2007) Thyroid hormone homeostasis and action in the type 2 deiodinase-deficient rodent brain during development. Endocrinology 148(7):3080–3088

Galton VA, Schneider MJ et al (2009) Life without thyroxine to 3,5,3′-triiodothyronine conversion: studies in mice devoid of the 5′-deiodinases. Endocrinology 150(6):2957–2963

Gan EH, Pearce SH (2012) Clinical review: the thyroid in mind: cognitive function and low thyrotropin in older people. J Clin Endocrinol Metab 97(10):3438–3449

Gil-Ibanez P, Morte B et al (2013) Role of thyroid hormone receptor subtypes alpha and beta on gene expression in the cerebral cortex and striatum of postnatal mice. Endocrinology 154(5):1940–1947

Grijota-Martinez C, Diez D et al (2011) Lack of action of exogenously administered T3 on the fetal rat brain despite expression of the monocarboxylate transporter 8. Endocrinology 152(4):1713–1721

Guadano-Ferraz A, Benavides-Piccione R et al (2003) Lack of thyroid hormone receptor alpha1 is associated with selective alterations in behavior and hippocampal circuits. Mol Psychiatry 8(1):30–38

Hadjab-Lallemend S, Wallis K et al (2010) A mutant thyroid hormone receptor alpha1 alters hippocampal circuitry and reduces seizure susceptibility in mice. Neuropharmacology 58(7):1130–1139

Hage MP, Azar ST (2012) The link between thyroid function and depression. J Thyroid Res 2012:590648

Hashimoto K, Curty FH et al (2001) An unliganded thyroid hormone receptor causes severe neurological dysfunction. Proc Natl Acad Sci U S A 98(7):3998–4003

Heinrich TW, Grahm G (2003) Hypothyroidism presenting as psychosis: myxedema madness revisited. Prim Care Companion J Clin Psychiatry 5(6):260–266

Hernandez A, Quignodon L et al (2010) Type 3 deiodinase deficiency causes spatial and temporal alterations in brain T3 signaling that are dissociated from serum thyroid hormone levels. Endocrinology 151(11):5550–5558

Hernandez A, Morte B et al (2012) Critical role of types 2 and 3 deiodinases in the negative regulation of gene expression by T3 in the mouse cerebral cortex. Endocrinology 153(6):2919–2928

Herwig A, Wilson D et al (2009) Photoperiod and acute energy deficits interact on components of the thyroid hormone system in hypothalamic tanycytes of the Siberian hamster. Am J Physiol Regul Integr Comp Physiol 296(5):R1307–R1315

Herwig A, de Vries EM et al (2013) Hypothalamic ventricular ependymal thyroid hormone deiodinases are an important element of circannual timing in the Siberian hamster (Phodopus sungorus). PLoS One 8(4):e62003

Heuer H, Mason CA (2003) Thyroid hormone induces cerebellar Purkinje cell dendritic development via the thyroid hormone receptor alpha1. J Neurosci 23(33):10604–10612

Heuer H, Visser TJ (2013) The pathophysiological consequences of thyroid hormone transporter deficiencies: insights from mouse models. Biochim Biophys Acta 1830(7):3974–3978

Howdeshell KL (2002) A model of the development of the brain as a construct of the thyroid system. Environ Health Perspect 110(Suppl 3):337–348

Kapoor R, Ghosh H et al (2011) Loss of thyroid hormone receptor beta is associated with increased progenitor proliferation and NeuroD positive cell number in the adult hippocampus. Neurosci Lett 487(2):199–203

Kinne A, Kleinau G et al (2010) Essential molecular determinants for thyroid hormone transport and first structural implications for monocarboxylate transporter 8. J Biol Chem 285(36):28054–28063

Koibuchi N (2013) The role of thyroid hormone on functional organization in the cerebellum. Cerebellum 12(3):304–306

Lang MF, Salinin S et al (2011) A transgenic approach to identify thyroxine transporter-expressing structures in brain development. J Neuroendocrinol 23(12):1194–1203

Lauder JM, Altman J et al (1974) Some mechanisms of cerebellar foliation: effects of early hypo- and hyperthyroidism. Brain Res 76(1):33–40

Lemkine GF, Raj A et al (2005) Adult neural stem cell cycling in vivo requires thyroid hormone and its alpha receptor. FASEB J 19(7):863–865

Liao XH, Di Cosmo C et al (2011) Distinct roles of deiodinases on the phenotype of Mct8 defect: a comparison of eight different mouse genotypes. Endocrinology 152(3):1180–1191

Lopez M, Varela L et al (2010) Hypothalamic AMPK and fatty acid metabolism mediate thyroid regulation of energy balance. Nat Med 16(9):1001–1008

Lopez-Juarez A, Remaud S et al (2012) Thyroid hormone signaling acts as a neurogenic switch by repressing Sox2 in the adult neural stem cell niche. Cell Stem Cell 10(5):531–543

Mayerl S, Visser TJ et al (2012) Impact of Oatp1c1 deficiency on thyroid hormone metabolism and action in the mouse brain. Endocrinology 153(3):1528–1537

McDonald MP, Wong R et al (1998) Hyperactivity and learning deficits in transgenic mice bearing a human mutant thyroid hormone beta1 receptor gene [In Process Citation]. Learn Mem 5(4–5):289–301

Mendes-de-Aguiar CB, Alchini R et al (2008) Thyroid hormone increases astrocytic glutamate uptake and protects astrocytes and neurons against glutamate toxicity. J Neurosci Res 86(14):3117–3125

Mittag J, Davis B et al (2010) Adaptations of the autonomous nervous system controlling heart rate are impaired by a mutant thyroid hormone receptor-alpha1. Endocrinology 151(5):2388–2395

Mittag J, Lyons DJ et al (2013) Thyroid hormone is required for hypothalamic neurons regulating cardiovascular functions. J Clin Invest 123(1):509–516

Montero-Pedrazuela A, Bernal J et al (2003) Divergent expression of type 2 deiodinase and the putative thyroxine-binding protein p29, in rat brain, suggests that they are functionally unrelated proteins. Endocrinology 144(3):1045–1052

Montero-Pedrazuela A, Venero C et al (2006) Modulation of adult hippocampal neurogenesis by thyroid hormones: implications in depressive-like behavior. Mol Psychiatry 11(4):361–371

Moran C, Schoenmakers N et al (2013) An adult female with resistance to thyroid hormone mediated by defective thyroid hormone receptor alpha. J Clin Endocrinol Metab 98(11):4254–4261

Morte B, Manzano J et al (2002) Deletion of the thyroid hormone receptor alpha 1 prevents the structural alterations of the cerebellum induced by hypothyroidism. Proc Natl Acad Sci U S A 99(6):3985–3989

Morte B, Diez D et al (2010) Thyroid hormone regulation of gene expression in the developing rat fetal cerebral cortex: prominent role of the Ca2+/calmodulin-dependent protein kinase IV pathway. Endocrinology 151(2):810–820

Navarro D, Alvarado M et al (2014) Late maternal hypothyroidism alters the expression of Camk4 in neocortical subplate neurons: a comparison with Nurr1 labeling. Cereb Cortex 24(10):2694–2706

Palha JA, Fernandes R et al (2000) Transthyretin regulates thyroid hormone levels in the choroid plexus, but not in the brain parenchyma: study in a transthyretin-null mouse model. Endocrinology 141(9):3267–3272

Peeters RP, Hernandez A et al (2013) Cerebellar abnormalities in mice lacking type 3 deiodinase and partial reversal of phenotype by deletion of thyroid hormone receptor alpha1. Endocrinology 154(1):550–561

Picou F, Fauquier T et al (2012) A bimodal influence of thyroid hormone on cerebellum oligodendrocyte differentiation. Mol Endocrinol 26(4):608–618

Pilhatsch M, Winter C et al (2010) Increased depressive behaviour in mice harboring the mutant thyroid hormone receptor alpha 1. Behav Brain Res 214(2):187–192

Pinna G, Brodel O et al (2002) Concentrations of seven iodothyronine metabolites in brain regions and the liver of the adult rat. Endocrinology 143(5):1789–1800

Portella AC, Carvalho F et al (2010) Thyroid hormone receptor beta mutation causes severe impairment of cerebellar development. Mol Cell Neurosci 44(1):68–77

Quignodon L, Legrand C et al (2004) Thyroid hormone signaling is highly heterogeneous during pre- and postnatal brain development. J Mol Endocrinol 33(2):467–476

Sait Gonen M, Kisakol G et al (2004) Assessment of anxiety in subclinical thyroid disorders. Endocr J 51(3):311–315

Sanchez E, Vargas MA et al (2009) Tanycyte pyroglutamyl peptidase II contributes to regulation of the hypothalamic-pituitary-thyroid axis through glial-axonal associations in the median eminence. Endocrinology 150(5):2283–2291

Schweizer U, Kohrle J (2013) Function of thyroid hormone transporters in the central nervous system. Biochim Biophys Acta 1830(7):3965–3973

Siesser WB, Zhao J et al (2006) Transgenic mice expressing a human mutant beta1 thyroid receptor are hyperactive, impulsive, and inattentive. Genes Brain Behav 5(3):282–297

Sousa JC, de Escobar GM et al (2005) Transthyretin is not necessary for thyroid hormone metabolism in conditions of increased hormone demand. J Endocrinol 187(2):257–266

Suzuki S, Suzuki N et al (2007) micro-Crystallin as an intracellular 3,5,3′-triiodothyronine holder in vivo. Mol Endocrinol 21(4):885–894

Tanti A, Belzung C (2013) Hippocampal neurogenesis: a biomarker for depression or antidepressant effects? Methodological considerations and perspectives for future research. Cell Tissue Res 354(1):203–219

Tinnikov A, Nordstrom K et al (2002) Retardation of post-natal development caused by a negatively acting thyroid hormone receptor alpha1. EMBO J 21(19):5079–5087

Trajkovic M, Visser TJ et al (2007) Abnormal thyroid hormone metabolism in mice lacking the monocarboxylate transporter 8. J Clin Invest 117(3):627–635

Vallortigara J, Chassande O et al (2009) Thyroid hormone receptor alpha plays an essential role in the normalisation of adult-onset hypothyroidism-related hypoexpression of synaptic plasticity target genes in striatum. J Neuroendocrinol 21(1):49–56

van Mullem A, van Heerebeek R et al (2012) Clinical phenotype and mutant TRα1. N Engl J Med 366(15):1451–1453

van Mullem AA, Chrysis D et al (2013) Clinical phenotype of a new type of thyroid hormone resistance caused by a mutation of the TRalpha1 receptor: consequences of LT4 treatment. J Clin Endocrinol Metab 98(7):3029–3038

Varela L, Martinez-Sanchez N et al (2012) Hypothalamic mTOR pathway mediates thyroid hormone-induced hyperphagia in hyperthyroidism. J Pathol 227(2):209–222

Vasudevan N, Morgan M et al (2013) Distinct behavioral phenotypes in male mice lacking the thyroid hormone receptor alpha1 or beta isoforms. Horm Behav 63(5):742–751

Venero C, Guadano-Ferraz A et al (2005) Anxiety, memory impairment, and locomotor dysfunction caused by a mutant thyroid hormone receptor {alpha}1 can be ameliorated by T3 treatment. Genes Dev 19(18):2152–2163

Wallis K, Dudazy S et al (2010) The thyroid hormone receptor alpha1 protein is expressed in embryonic postmitotic neurons and persists in most adult neurons. Mol Endocrinol 24(10):1904–1916

Wilcoxon JS, Nadolski GJ et al (2007) Behavioral inhibition and impaired spatial learning and memory in hypothyroid mice lacking thyroid hormone receptor alpha. Behav Brain Res 177(1):109–116

Williams GR (2008) Neurodevelopmental and neurophysiological actions of thyroid hormone. J Neuroendocrinol 20(6):784–794

Wirth EK, Roth S et al (2009) Neuronal 3′,3,5-triiodothyronine (T3) uptake and behavioral phenotype of mice deficient in Mct8, the neuronal T3 transporter mutated in Allan-Herndon-Dudley syndrome. J Neurosci 29(30):9439–9449

Yilmazer-Hanke DM, Hantsch M et al (2004) Neonatal thyroxine treatment: changes in the number of corticotropin-releasing-factor (CRF) and neuropeptide Y (NPY) containing neurons and density of tyrosine hydroxylase positive fibers (TH) in the amygdala correlate with anxiety-related behavior of Wistar rats. Neuroscience 124(2):283–297

Yoshimura T (2013) Thyroid hormone and seasonal regulation of reproduction. Front Neuroendocrinol 34(3):157–166

Chapter 9
Disruption of Auditory Function by Thyroid Hormone Receptor Mutations

David S. Sharlin

Abstract More than a century ago, associations between deaf-mutism, cretinism, and goiter were reported. Since then, our understanding for the role of T3 and its receptors in auditory system development has greatly advanced. Much of current understanding is due to the creation of genetic mouse models that have allowed for the dissection of individual components of thyroid hormone action, including the thyroid hormone receptors. This chapter highlights our existing knowledge for the role of T3 and the thyroid hormone receptors, TRα and TRβ, in cochlear development and auditory function that have obtained from these important genetic models. It also discusses auditory deficits in humans with the syndrome of resistance to thyroid hormone caused by mutations in the thyroid hormone receptors. It concludes by briefly discussing where the deficiencies in our understanding of thyroid hormone action in cochlear development remain.

Keywords Thyroid hormone • Thyroid hormone receptor • Cochlea • Hearing • Development • Inner ear

9.1 Introduction

One of the most prominent consequences of congenital thyroid diseases is profound deafness. Clinical reports dating as far back as the late 1800s, well before the identification of the thyroid hormones, described associations between deaf-mutism, cretinism, and goiter. Since then, our mechanistic and functional understanding for role of thyroid hormone and thyroid hormone receptors in the development of the auditory system have greatly advanced (Ng et al. 2013). Much of our understanding is rooted in correlating functional studies in humans with thyroid disorders with

D.S. Sharlin (✉)
Department of Biological Sciences, Minnesota State University, Mankato,
242 Trafton Science Center South, Mankato, MN 56001, USA
e-mail: david.sharlin@mnsu.edu

© Springer Science+Business Media New York 2016
N. Koibuchi, P.M. Yen (eds.), *Thyroid Hormone Disruption and Neurodevelopment*,
Contemporary Clinical Neuroscience, DOI 10.1007/978-1-4939-3737-0_9

those conducted in model species such as genetically engineered mice. These studies have led investigators to propose that a key function for the main active thyroid hormone (T3; triiodothyronine) in auditory development is to synchronize the maturation of cochlear tissues at a developmental time that is permissive for the onset of hearing (Ng et al. 2013).

These developmental events range from cochlear tissue remodeling events to the regulation of ion channels to the physiological function of individual hair cells. Through these studies, it has become increasingly clear that action of thyroid hormone at the level of the receptor (and thus developmental events they control) is regulated not only by circulating levels of the thyroid hormones, but also by upstream mediators of thyroid hormone action such as the deiodinase enzymes and potentially thyroid hormone membrane transporters (Visser et al. 2008; Arrojo et al. 2013).

The purpose of this chapter is to examine our current understanding of the role of thyroid hormone receptors in auditory development and function. It begins by examining the basic structure of the auditory system to provide essential background. It then discusses deficits in auditory processing and defects in cochlear anatomy observed with insufficient thyroid hormone during development as mode to inform the phenotypes observed with thyroid hormone receptor mutations. The chapter then highlights our current knowledge of the roles of thyroid hormone receptor alpha and beta (TRα and TRβ) in the development of auditory function, cochlear anatomy, and hair cell physiology and attempts to link defects observed with low thyroid hormone with receptor-specific functions. Lastly, this chapter concludes by briefly considering where our knowledge is lacking with a hope to stimulate further research in the field.

9.2   The Auditory System

The mammalian auditory system can be divided into three anatomical regions—the outer ear, middle ear, and inner ear. The outer ear functions to collect airborne sound waves from the pinna and outer ear canal (auditory meatus) to the tympanic membrane (ear drum). Collected sound waves result in vibration of the tympanic membrane. The resulting tympanic membrane vibrations are distributed and amplified through the three middle ear bones known as the ossicular chain (malleus, incus, stapes) located within the air-filled middle ear cavity. The most lateral of the three bones, the stapes, sits in the oval window of the cochlea and converts mechanical vibrations to pressure wave inside the fluid-filled cochlea contained within the inner ear. Pressure wave travels along the cochlear spiral and activates hairs in a frequency-specific manner (McKinley and O'Loughlin 2012).

Hair cells are contained within a specialized epithelium, the Organ of Corti, which runs the length of the cochlear duct. Within the organ of Corti, a single row of inner hairs and three rows of outer hairs detect and refine auditory output form the cochlea, respectively (Kelley 2006). The acellular tectorial membrane that attaches to the superior surface of hairs is responsible for deflecting stererocilia when the elastic basilar membrane is induced to vibrate in response to the pressure waves (Lukashkin et al. 2010). Deflection of stereocilia results in the opening of mechanically gated K^+ channels which results in hair cell depolarization and release of neurotransmitter from the base of hair cell on to the peripheral processes of afferent neurons located in the spiral ganglion. Hair cell depolarization is dependent on a specialized, K^+-rich fluid known as endolymph contained within the cochlear duct (scala media). This high concentration of K^+ generates the endocochlear potential, the main driving force for sensory transduction (Wangemann 2006). Action potentials generated in the spiral ganglion neurons are passed via the axons to the brainstem and central auditory pathways (Rubel and Fritzsch 2002).

9.3   Thyroid Hormone and Auditory System Development

A key function of thyroid hormone is to stimulate development of the auditory system. In humans, deafness is well known to be associated with developmental hypothyroidism including congenital hypothyroidism (CH) (DeLong et al. 1985; Rovet et al. 1996). Moreover, congenitally hypothyroid CH individuals identified by neonatal screening and adequately treated were significantly more likely (three times) than their peers to report self-declared hearing loss (Leger et al. 2011). Audiologic analysis in this population confirmed hearing loss and indicated that in the majority of CH patients with hearing loss, it was bilateral, concerned mild to moderate loss of higher frequencies, and of sensorineural type (Lichtenberger-Geslin et al. 2013). Although implementation of neonatal screen for CH has greatly reduced the severity of neurological deficits (Zoeller and Rovet 2004), including hearing (Lichtenberger-Geslin et al. 2013), residual hearing loss remains a significant problem in CH patients.

Experimental rodent models have indicated that thyroid hormone has a role in the maturation of most regions of the auditory system. In congenitally hypothyroid animals, a delayed raising of the pinnae is observed, indicating a T3 signal is required for proper timing of outer ear maturation that occurs during the first 2 postnatal weeks (Sprenkle et al. 2001a, b). Concomitant with outer ear maturation is the cavitation of the middle ear cavity and clearance of middle ear mesenchyme that surrounds the ossicular chain (Mallo 2001). Delayed clearance of middle ear mesenchyme and altered ossicular bone development is reported in developmentally hypothyroid animals as well as $Thra^{(+/PV)}$ mice that express a dominant-negative $TR\alpha1$ protein that results in a cellular "hypothyroidism" (Cordas et al. 2012; Fraser 1932).

During cochlear development, a major role for thyroid hormone is to mediate the remodeling of the greater epithelial ridge (GER; also termed Kölliker's organ) (Deol 1976), a transient epithelial cell structure located medially to the sensory hair cells (Fig. 9.1). Although the functions of the GER are not fully understood, it is accepted that GER cells secrete several important sulfonated glycoproteins that

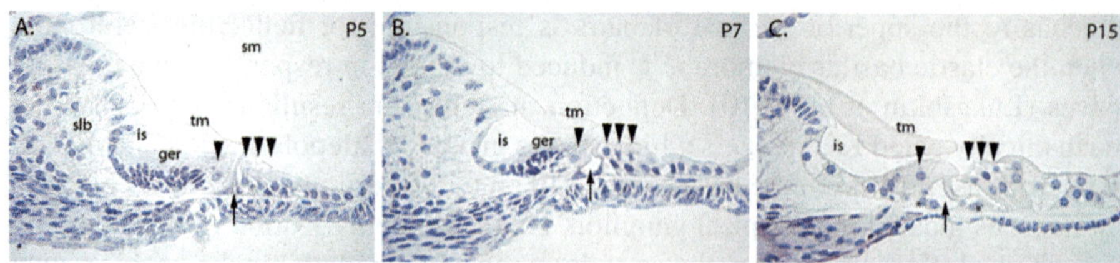

Fig. 9.1 Postnatal remodeling in the rodent cochlea. Developmental ages depicted are P5 (**a**), P7 (**b**), and P15 (**c**). Each image shows a mid-basal turn of the cochlea. Two major remodeling events are histologically evident during this 2-week period. (1) Regression of the greater epithelial ridge (ger) forming the inner sulcus (is) and (2) opening of the tunnel of corti (indicated by the *arrow*). *Arrowheads* indicate the single inner row of hair cells and three rows of outer hair cells. Cells of greater epithelial ridge that secrete extracellular matrix proteins to form the tectorial membrane (tm) slowly disappear as development proceeds. The tunnel of corti, a triangular shaped fluid-filled space, separates the inner and outer hair cells and is dependent on pillar cell maturation (*asterix*). Note that the tectorial membrane (tm) contacts the hair cells. *slb* spiral limbus, *sm* scala media

contribute to the formation of the tectorial membrane (Legan et al. 1997; Rau et al. 1999; Killick and Richardson 1997). More recently, GER cells were identified as a source of ATP that induces spontaneous electrical activity in inner hair cells that is required for neuronal survival and refinement of the tonotopic map in the brain (Tritsch et al. 2007; Leao et al. 2006). In euthyroid rodents, the GER regresses during the first 2 weeks of postnatal development before the onset of hearing creating the inner sulcus that allows proper suspension of the tectorial membrane (Hinojosa 1977; Fig. 9.1).

Although the timing of GER remodeling occurs in parallel with a natural increase in circulating levels of thyroid hormones in the blood, several studies have demonstrated that the remodeling is controlled by both systemic and local control mechanisms. Ng and colleagues (2009) demonstrated that prior to birth and shortly thereafter, activation of thyroid hormone receptors locally in the cochlea was protected from the increasing levels of serum T3 by inactivating type 3 deiodinase (Dio3). Furthermore, these authors reported that Dio3 is expressed in the GER and that deletion of Dio3 resulted in premature GER remodeling, a thin tectorial membrane, and permanent auditory deficits. Additionally, these authors showed crossing Dio3 mutant mice with mice lacking TRβ resulted in phenotype similar to that of TRβ alone; providing evidence that premature GER remodeling was mediated through the TRβ receptor.

Shortly after birth, Dio3 is robustly down-regulated in the cochlea and the activating type 2 deiodinase (Dio2) is significantly up-regulated (Campos-Barros et al. 2000; Ng et al. 2009). In mice lacking Dio2, GER remodeling is delayed and the mice have permanent auditory deficits (Ng et al. 2004). This observation is similar to developmentally hypothyroid animals (Deol 1976; Mustapha et al. 2009; Sharlin et al. 2011; Uziel et al. 1981; Fig. 9.2) and suggests that Dio2 is required to locally amplify the local T3 pool needed for GER remodeling. Delayed GER remodeling in hypothyroid animals is associated with a deformed tectorial membrane (Deol 1976; Uziel et al. 1981, 1983; Fig. 9.2) and is likely responsible for some of the functional

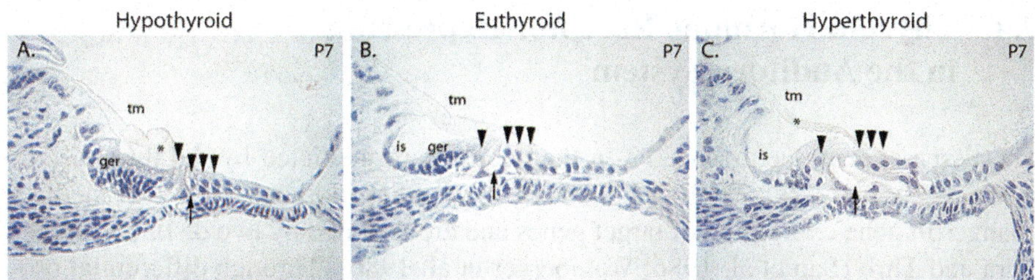

Fig. 9.2 T3 and postnatal remodeling in the rodent cochlea. Each image shows a mid-basal turn of the cochlea at postnatal day 7 (P7) from a hypothyroid (**a**), euthyroid (**b**), or hypothyroid (**c**) animal. Note that the regression of greater epithelial ridge cells (ger) is delayed in the hypothyroid cochlea (**a**) and advanced in the hyperthyroid cochlea (**c**) compared to euthyroid control animals (**b**). This delay or advancement in greater epithelial ridge cell regression results in an enlarged (swollen) tectorial membrane (tm) or thin tectorial membrane, respectively (*asterix* in (**a**, **c**)). In addition, opening of the tunnel of corti (*arrow*) is delayed in the hypothyroid cochlea and advanced in the hyperthyroid cochlea compared to the euthyroid cochlea. Note that under either hypothyroid or hyperthyroid conditions, cochlea hair cells (*arrowheads*) are present in correct numbers at this developmental age. *is* inner sulcus

hearing deficits observed in hypothyroid animals (Sprenkle et al. 2001b; Uziel et al. 1980). It should be noted that the mechanism (i.e., TR target genes) by which thyroid hormone regulates the timing and regression of GER cells is completely unknown and most certainly warrants future investigations.

Another structural maturation in the cochlea is opening of the tunnel of corti. Inner and outer hair cells are separated by a specialized pair of cells known as the pillar cells that run the length of the cochlear duct. This separation is called as the tunnel of corti and consists of triangular fluid-filled space (Raphael and Altschuler 2003). Around the end of the first postnatal week, the tunnel of corti begins to open and is fully expanded by the onset of hearing. Developmental hypothyroidism delays tunnel of corti opening and permanently reduces pillar cell height (Uziel et al. 1983). These observed effects may be attributed to thyroid hormone regulating Fgf3 signaling in pillar cells and altered cytoskeletal dynamics (Szarama et al. 2013). However, TR expression in pillar cells has not been clearly documented. Therefore, it is currently unknown whether the effects of hypothyroidism on tunnel of corti opening are mediated directly in pillar cells or indirect through neighboring tissues that are TR positive.

In addition to role of T3 mediating GER remodeling and opening of the tunnel of corti, several other developmental events are altered following insufficient thyroid hormone. Knipper and colleagues (1998) demonstrated that myelination of the vestibulocochlear nerve (cranial nerve VIII), which transmits electrical signals from the spiral ganglion neurons to the brain, is delayed in hypothyroid animals. Mustapha et al. (2009) reported altered stria vascularis structure and low endocochlear potential in the Pit1$^{dw/dw}$ hypothyroid model. Furthermore, these authors demonstrated permanently reduced potassium channel expression that likely contributes to physiological and functional defects detected in these hypothyroid animals.

9.4 Thyroid Hormone Receptor Expression in the Auditory System

The tissue-specific actions of T3 in the cochlea are mediated by the thyroid hormone receptors (TRs). TRs are ligand-regulated transcription factors that mediate changes in gene expression of target genes and are encoded by two distinct genes— Thra and Thrb (Sap et al. 1986; Weinberger et al. 1986). Through differential promoter usage and alternative splicing, these two genes produce 4 major receptor isoforms in the mouse: TRα1, TRα2, TRβ1, and TRβ2 (Zhang and Lazar 2000; Jones et al. 2003). TRα1, TRβ1, and TRβ2 are bona fide receptors that bind T3 with nearly identical affinities (Schwartz et al. 1992; Oppenheimer et al. 1994). TRα2 binds DNA, but contains an alternatively spliced C-terminal that changes the ligand-binding domain amino acid sequence and is unable to bind T3 (Lazar et al. 1988). Therefore, TRα2 is not considered as bona fide thyroid hormone receptor.

Analysis of TR mRNA expression in the cochlea indicated developmental and cell-specific expression profiles. One of most comprehensive analyses of TR expression was completed by Bradley et al. (1994). Using in situ hybridization this report indicated that all four major TR isoforms are expressed in the cochlea. Temporally, in contrast to TR expression in the central nervous system where TRα is the predominantly expressed isoform during development (Bradley et al. 1992), TRβ is the predominant isoform expressed in the developing cochlea with TRβ1 expression greater than TRβ2. TRβ1 and TRβ2 isoforms have overlapping expression in several cochlear structures including the greater and lesser epithelial ridges, outer sulcus epithelium, spiral limbus, and sensory epithelium with lower levels noted in the spiral ganglion (Bradley et al. 1994; Fig. 9.3a, b). Peak TRβ mRNA expression in the cochlea occurs shortly after birth and falls slightly after the first postnatal week

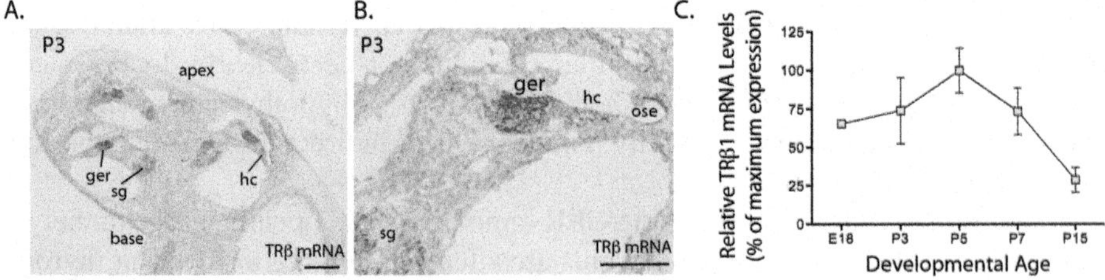

Fig. 9.3 Expression of thyroid hormone receptor beta (TRβ) mRNA in the perinatal cochlea. (**a**) Expression of TRβ mRNA in the postnatal day 3 (P3) cochlea visualized by in situ hybridization. This low power image shows three cochlea turns. TRβ expression is prominent in the greater epithelial ridge (ger) and spiral ganglion (sg). (**b**) Under high power magnification, TRβ expression can also be visualized in the sensory epithelium that contains the hair cells (hc) and outer sulcus epithelium (ose). (**c**) Relative levels of TRβ mRNA during cochlear development. Data are normalized to maximal expression detected at P5. Each point represents mean ± SEM for three individual cochleae. The analysis reveals the trend in TRβ expression levels over the developmental time indicated. Scale bar = 100 µm. Figure is adopted from (Sharlin et al. 2011)

(Sharlin et al. 2011; Fig. 9.3c). In contrast, TRα1 expression is comparatively lower in the greater and lesser epithelial ridges, outer sulcus epithelium, spiral limbus, and sensory epithelium than TRβ1 and TRβ2, but higher in the spiral ganglia. The expression of TRα2 in the cochlea was very similar to the expression pattern of TRα1. However, TRα2, in contrast to TRα1, was also noted within the vestibular complex (Bradley et al. 1994). Largely consistent with mRNA studies, immunohistochemical studies have localized TRα and TRβ proteins to hair cells, but confirmation in other cochlear tissues is lacking (Knipper et al. 1999; Lautermann and ten Cate 1997). Outside the cochlea, both TRα and TRβ expression have been documented in the middle ear (Cordas et al. 2012) and central auditory pathway (Bradley et al. 1992).

9.5 Targeted Thyroid Hormone Receptor Deletions in Mice

9.5.1 Contribution of TRβ and TRα in Auditory Function

Measurement of auditory-evoked brainstem response thresholds (ABR) and distortion product otoacoustic emissions (DPOAE), functional hearing analyses that test sound coding by inner hair cells and outer hair cells function, respectively, have demonstrated specific roles for TRβ and TRα in the development of auditory function. In mice lacking both TRβ1 and TRβ2 (denoted as TRβ), ABR thresholds are permanently elevated across a wide range of frequencies (Forrest et al. 1996; Winter et al. 2009); a phenotype consistent with developmental hypothyroidism (Uziel et al. 1980). Analysis of DPOAE amplitudes and thresholds demonstrated that in absence of TRβ, active cochlear mechanics are greatly inhibited at frequencies above 8 kHz (Winter et al. 2009). Auditory function, as assessed by ABR, was normal in mice with a TRβ2 isoform-specific deletion (Abel et al. 1999); suggesting that TRβ1 is the major TRβ isoform mediating thyroid hormone action in the developing cochlea. Furthermore, taken together, these observations suggest that although TRβ and TRα are both expressed in the cochlea with partial overlap (Bradley et al. 1994), TRα cannot completely compensate for the loss of TRβ in the development of auditory function.

Mice deficient in TRα1 have normal ABR thresholds (Rusch et al. 1998), suggesting that this receptor isoform is individually dispensable for the development of functional hearing as measured by ABR. An ancillary role for TRα1 in auditory function was unmasked by generating mice lacking all known thyroid hormone receptors. Mice lacking TRβ and TRα1 have ABR thresholds that are 20 dB higher than mice lacking solely TRβ (Rusch et al. 2001). This result suggests that, at least in part, a subset of developmental programs regulated by thyroid hormone receptors in the cochlea and required for normal hearing overlap between the two thyroid hormone receptor gene products.

Similar to mice lacking TRα1, Ng et al. (2001) demonstrated that in mice that lack TRα2—a TR isoform described as a TR antagonist that neither binds thyroid hormone nor transactivates (Koenig et al. 1989; Lazar et al. 1988)—and concomitantly overexpress TRα1, have normal ABR thresholds indicating that TRα2 is dispensable. Interestingly, crossing this strain with TRβ-null mice rescues the auditory phenotype observed in TRβ-null (Ng et al. 2001). It was hypothesized that the overexpression of TRα1 substitutes for the loss of TRβ. This idea is supported by the observation that mice heterozygous for the TRα2 mutation can rescue TRβ-null phenotype. Mice lacking all TRα isoforms (TRα1 and TRα2) were reported to have normal hearing at frequencies between 4 and 16 kHz, but elevated thresholds at 24 kHz (Gauthier et al. 2001). The relationship between TRα function and frequency-dependent hearing is unknown, but such frequency-specific defects have not been reported in mice harboring dominant negative TRα proteins (Cordas et al. 2012; Dettling et al. 2014).

9.5.2 Contribution of TRβ and TRα in Cochlear Anatomy

To understand the functional deafness reported in mice with target thyroid hormone receptor mutations, extensive histological and physiological analyses have been completed. As indicated earlier, a T3 signal is required for the proper timing of GER remodeling and tunnel of corti opening. Inactivation of TRβ results in delayed remodeling of the greater epithelial ridge that is accompanied by malformation of the tectorial membrane exhibiting a swollen presentation and ultrastructural disarray of protein fibers (Rusch et al. 2001; Winter et al. 2009); observations consistent with a hypothyroid phenotype (Deol 1973; Uziel et al. 1981). This malformation likely results, at least in part, from perturbing the temporal regulation of tectorin production in the GER (Rau et al. 1999); which would result in an altered production of key proteinaceous components of the tectorial membrane such as the tectorins (Knipper et al. 2001; Rusch et al. 2001; Winter et al. 2009). Tectorins are glycoproteins that represent major components of the tectorial membrane (Legan et al. 1997) and are essential for normal hearing in mice and humans (Legan et al. 2000; Verhoeven et al. 1998; Mustapha et al. 1999). Therefore, it is reasonable to speculate that a deformed tectorial membrane would alter deflection of hair cell stereocilia contributing to the auditory dysfunction observed in TRβ knockout mice. The presence of elevated ABR thresholds and reduced DPOAE amplitudes discussed earlier is consistent with the idea of hair cell dysfunction from improper stereocilia deflection. Furthermore, considering that GER cells also release adenosine triphosphate (ATP) that initiates spontaneous electrical activity of inner hair cells prior to the onset of hearing (Tritsch and Bergles 2010; Tritsch et al. 2007) is attractive to speculate that sustained ATP release from the residual GER cells might alter inner hair cell function resulting in an altered tonotopic map.

9.5.3 Contribution of TRβ and TRα to Hair Cell Function

In addition to the role of TRβ in GER remodeling and tectorial membrane formation, physiological investigation has demonstrated a role for TRs in hair cell function. Inner hair cells transform sound-induced mechanical stimulation into graded receptor potentials (membrane depolarization) that results in neurotransmitter release and excitation of afferent auditory nerve fibers (Gillespie and Muller 2009). Repolarization of the membrane potential requires a fast-activating potassium current, $I_{k,f}$ that is first observed around postnatal day 13 and is associated with inner hair cell maturation and the onset of hearing (Kros et al. 1998). The $I_{k,f}$ currents are carried by large conductance, voltage- and Ca^{++}-activated K (BK) channels and mature reaching a maximal inward potassium current around postnatal day 20 (Pyott et al. 2007; Kros et al. 1998). In mice lacking TRβ, $I_{k,f}$ currents were greatly diminished between P15-P18 compared to controls and remained reduced until about P40 when $I_{k,f}$ magnitudes normalized and reached those seen in wild-type animals (Rusch et al. 1998). It was proposed that the delayed $I_{k,f}$ might alter maturation of auditory brainstem neurons contributing to the permanent deafness observed in TRβ knockouts. However, since this report, mice lacking a major component of the BK channel have been shown to have normal hearing (Pyott et al. 2007); suggesting that TRβ controls the timing ionic conductance in inner hair cells, but that the permanent deafness is not likely the result of this observed delay.

Two recent studies (Dettling et al. 2014; Winter et al. 2009) have utilized a conditional mutagenesis approach to understand the roles of TRβ specifically in hair cells. Using either a prestin-cre or math1-creER™ transgenic mice, TRβ was conditionally deleted in hair cells (discussed as TRβ-hc) at around P11 or embryonic day 14 (E14), respectively. In both these mouse experiments, auditory function, as measured by ABR, was largely normal (Dettling et al. 2014; Winter et al. 2009). In addition, loss of TRβ in prestin-positive hair cells resulted in a delayed $I_{k,f}$ similar to that of constitutive TRβ mutants (Rusch et al. 1998). Consistent with these physiological measurements , expression of BKα, the pore-forming subunit, was transiently reduced in inner hair cells between P15 and P30 (Winter et al. 2009). This finding suggests that the delayed $I_{k,f}$ observed in TRβ mutants is due to a direct role of TRβ in the inner hair cells as surrounding tissue expression of TRβ should be intact. When taken together, it appears as though profound deafness observed in TRβ knockout mice originates from outside the hair cells.

Measurements of compound action potentials (CAP), which are the synchronous activity of all the individual action potentials after a sound stimulus, provide information on electrical activity emanating from the cochlea. CAP thresholds in TRβ knockouts were elevated across a wide range of frequencies. Although the maximum CAP amplitudes and waveform were similar between TRβ knockouts and wild-type controls, CAP input–output functions of TRβ mutants were shifted to higher SPLs with steeper slopes (Winter et al. 2009). Poor mechanical excitation of hair cells, perhaps due to the defects in cochlear remodeling and tectorial membrane reported (Rusch et al. 2001; Winter et al. 2009), could explain these observations.

Although TRα1 and TRα2 are independently or combinatorial dispensable for the development of auditory function (Ng et al. 2001; Rusch et al. 1998, 2001), several reports indicate that TRα1 has a role in regulating ion channel protein level during outer hair cell differentiation potentially altering hair cell function. In outer hair cells, three different K^+ channels, Kcnq4, SK2, and BK have retarded expression under hypothyroid conditions that is reversed following deletion of TRα1 (Winter et al. 2006, 2007). This observation suggests that the unliganded TRα1 receptor mediates the observed repression. This finding is similar to the effect of unliganded TRα1 on cerebellar maturation (Morte et al. 2002).

9.6 Thyroid Hormone Receptor Mutations in Humans and Auditory Function

In humans, mutations in TRα (THRA) and TRβ (THRB) result in the syndrome of resistance to thyroid hormone (RTH) [reviewed (Dumitrescu and Refetoff 2013; Schoenmakers et al. 2013)]. Although these forms of RTH manifest with somewhat different clinical features, both are usually observed as autosomal dominant syndromes that are the consequence of heterozygous mutations that generate dominant-negative proteins (Sakurai et al. 1990; van Mullem et al. 2012; Bochukova et al. 2012).

Significant hearing deficits are observed in 21 % of RTH patients with heterozygous mutations in THRB. The reported hearing loss was characterized, with equal frequency, to be of the conductive and sensorineural types. DPOAE analysis implicated cochlear dysfunction in a subset of these patients (Brucker-Davis et al. 1996). RTH caused by rare homozygous mutations in THRB is associated with deaf-mutism (Ferrara et al. 2012; Refetoff et al. 1967). In contrast to the hearing loss reported in the subset of RTH patients with heterozygous mutations in THRB, mice carrying a heterozygous knock-in TRβ mutation lacks T3 binding and that mimics human RTH (TRβ^{PV}; Kaneshige et al. 2000; Parrilla et al. 1991), have normal hearing function (Griffith et al. 2002). However, mice homozygous for the TRβ^{PV} mutation have severe hearing loss that is accompanied by delayed GER remodeling and deformed tectorial membrane (Griffith et al. 2002). These reported phenotypes are similar to those reported for developmentally hypothyroid mice or mice lacking TRβ (Deol 1976; Mustapha et al. 2009; Forrest et al. 1996; Rusch et al. 2001; Winter et al. 2009).

Human mutations in the THRA gene were only recently reported (Bochukova et al. 2012; van Mullem et al. 2012). Although these patients have clinical features that are similar to hypothyroidism, congenital deafness is not present. One patient is reported to have acquired hearing loss due to otosclerosis, but this condition likely developed independent from the THRA mutation (van Mullem et al. 2013). The lack of an auditory phenotype in these patients would not have been predicted based on two reports that investigated auditory function in mice with dominant-negative

TRα receptors (Cordas et al. 2012; Dettling et al. 2014). Mice constitutively expressing a dominant-negative TRα^{PV} mutation (Kaneshige et al. 2001), which is a frameshift mutation resulting in truncated C-terminal similar to one of the mutations reported for THRA (van Mullem et al. 2012), were reported to have reduced ABR and DPOAE thresholds. These functional deficits were attributed to delayed mesenchyme disappearance and altered ossicular bone development (Cordas et al. 2012). These authors also reported that the effects were present in developmentally hypothyroid animals (Tshr knockouts), suggesting that functional auditory deficits reported for other hypothyroid models might be due to a combination of cochlear defects and conductive defects. Recently, Dettling et al. (2014) reported enhanced auditory output from inner and outer hairs after targeted expression of the dominant negative TRα1^{L400R} receptor to hair cells using a Math1-Cre. Although the mechanism behind the enhancement is unknown, the authors hypothesize that alterations in Ca^{++} currents might underlie this observation.

9.7 Concluding Remarks

This chapter has reviewed our current understanding of the role of thyroid hormone and thyroid hormone receptors in development and function of the auditory system. Largely through the use of mouse genetic models (see summary in Table 9.1), our understanding of the mechanistic and functional roles of thyroid hormone receptors in cochlear development has greatly advanced. These studies have demonstrated a major role for TRβ in the timing of cochlear maturation. Specifically, TRβ temporally regulates GER and tunnel of corti remodeling and mis-regulation of this timing is associated with hearing loss. Additionally, TRβ regulates inner hair cell physiology by controlling, in part, the expression of ion channel expression. TRα has a contributory role in cochlear development and function, but these genetic studies indicate TRα is dispensable.

Although much progress has been made in understanding the functions of thyroid hormone receptors in cochlea development, several unknowns remain. First, cochlear development and auditory function in mice lacking solely TRβ1 have not been reported. Considering that mice lacking TRβ2 have normal hearing (Abel et al. 1999) and that TRβ1 and TRβ2 have similar expression in the cochlea (Bradley et al. 1994), it is plausible that TRβ1 and TRβ2 may have overlapping functions. Investigating auditory function in mice lacking TRβ1 would address this question. Recently, the auditory phenotype of mice lacking solely TRβ1 was reported. Although a relatively mild developmental phenotype was observed, marked age-related hearing loss and hair cell degeneration was noted (Ng et al. 2015). Second, only a few candidate direct TR target genes in the cochlea have been reported (Weber et al. 2002; Winter et al. 2006). The genes identified, Kcnq4 and prestin, are expressed in hair cells and are important in hair cell physiology and electromotility, respectively. However, an important function of T3 action, as discussed in detail earlier, is to mediate GER remodeling. Kcnq4 and prestin are not expressed within

Table 9.1 Summary of auditory defects in mice with thyroid hormone receptor mutations

Gene	Targeted protein	Strategy	TR function	Cre line	Reported phenotype	Reference
Thrb	TRβ (Thrb[tm1])	Knockout	Absent	–	Hearing loss, delayed GER regression, malformed TM, delayed I_{kf} currents, altered prestin localization	Forrest et al. (1996), Rusch et al. (1998, 2001), and Winter et al. (2006, 2009)
	TRβ2	Knockout	Absent	–	Normal hearing	Abel et al. (1999)
	TRβ L2	Conditional	Dominant negative	Prestin-Cre; hair cell specific	Normal hearing, delayed BK channel expression, and I_{kf} currents	Winter et al. (2009)
	TRβ L2	Conditional	Dominant negative	Math1-CreER™; hair cell specific	Normal ABR, slightly enhance DPOA growth function, delayed BK channel expression	Dettling et al. (2014)
	TRβ (Thrb[PV])	Knockin	Dominant negative	–	Thrb[PV/+] normal. Thrb[PV/PV] severe hearing loss, malformed TM	Griffith et al. (2002)

Thra	TRα1 (Thra^{tm1})	Knockout	Absent	–	Normal hearing function, regulates K$^+$ channels	Rusch et al. (1998) and Winter et al. (2007)
	TRα (Thra$^{0/0}$)	Knockout	Absent	–	Normal hearing function	Gauthier et al. (2001)
	TRα2 (Thra^{tm2})	Knockout	Absent	–	Normal hearing function	Ng et al. (2001)
	TRα1^{AMI} (TRα1^{L400R})	Conditional	Dominant negative	SYCP1-Cre (sperm expressed)	Reduced Kcnq4 expression in outer hair cells	Quignodon et al. (2007)
	TRα1^{AMI} (TRα1^{L400R})	Conditional	Dominant negative	Math1-CreER™; hair cell specific	Enhanced ABR peak amplitudes, enhanced DPOAE function	Dettling et al. (2014)
	TRα1^{R384C}	Knockin	Dominant negative	–	Reduced Kcnq4 expression	Winter et al. (2006)
	TRα (Thrα^{PV})	Knockin	Dominant negative	–	Hearing loss, delayed middle ear mesenchyme clearance, enlarged ossicles	Cordas et al. (2012)
Thrb/Thra	Thrb^{tm1} × Thra^{tm1} double mutant	Knockout	Absent	–	Similar defects to Thrb^{tm1} but with overall enhanced severity	Rusch et al. (2001)
	Thrb^{tm1} × Thra^{tm2} double mutant	Knockout	Absent	–	Overexpression of TRα1 in Thra^{tm2} strain rescues Thrb^{tm1} phenotype	Ng et al. (2001)

the GER and therefore are not likely regulating GER regression. Understanding the transcription programs regulated by T3 that drive GER remodeling is an important and open area of research. It is intriguing to speculate that by determining the direct TR targets in the GER that are responsible for mediating regression of this structure could result in the identification of novel human deafness genes.

Lastly, we still have little understanding of the mechanisms that regulate thyroid hormone receptor activity during cochlear development. The deiodinase enzymes (Dio2 and Dio3) control thyroid hormone availability in the cochlea and are essential for normal cochlear development and auditory function (Ng et al. 2004, 2009). Considering that these enzymes show dynamic expression profiles during cochlear development, their transcriptional regulation is an important concept to consider. Recently, several membrane transporters that mediate the cellular uptake and efflux of thyroid hormones were identified in the developing cochlea with temporally and spatially specific expression patterns (Sharlin et al. 2011). However, whether thyroid hormone transporters are necessary for mediating thyroid hormone action during cochlear development is unknown and warrants investigation.

References

Abel ED, Boers ME, Pazos-Moura C, Moura E, Kaulbach H, Zakaria M, Lowell B, Radovick S, Liberman MC, Wondisford F (1999) Divergent roles for thyroid hormone receptor beta isoforms in the endocrine axis and auditory system. J Clin Invest 104(3):291–300. doi:10.1172/JCI6397

Arrojo EDR, Fonseca TL, Werneck-de-Castro JP, Bianco AC (2013) Role of the type 2 iodothyronine deiodinase (D2) in the control of thyroid hormone signaling. Biochim Biophys Acta 1830(7):3956–3964. doi:10.1016/j.bbagen.2012.08.019

Bochukova E, Schoenmakers N, Agostini M, Schoenmakers E, Rajanayagam O, Keogh JM, Henning E, Reinemund J, Gevers E, Sarri M, Downes K, Offiah A, Albanese A, Halsall D, Schwabe JW, Bain M, Lindley K, Muntoni F, Vargha-Khadem F, Dattani M, Farooqi IS, Gurnell M, Chatterjee K (2012) A mutation in the thyroid hormone receptor alpha gene. N Engl J Med 366(3):243–249. doi:10.1056/NEJMoa1110296

Bradley DJ, Towle HC, Young WS III (1992) Spatial and temporal expression of alpha- and beta-thyroid hormone receptor mRNAs, including the beta 2-subtype, in the developing mammalian nervous system. J Neurosci 12(6):2288–2302

Bradley DJ, Towle HC, Young WS III (1994) Alpha and beta thyroid hormone receptor (TR) gene expression during auditory neurogenesis: evidence for TR isoform-specific transcriptional regulation in vivo. Proc Natl Acad Sci U S A 91(2):439–443

Brucker-Davis F, Skarulis MC, Pikus A, Ishizawar D, Mastroianni MA, Koby M, Weintraub BD (1996) Prevalence and mechanisms of hearing loss in patients with resistance to thyroid hormone. J Clin Endocrinol Metab 81(8):2768–2772. doi:10.1210/jcem.81.8.8768826

Campos-Barros A, Amma LL, Faris JS, Shailam R, Kelley MW, Forrest D (2000) Type 2 iodothyronine deiodinase expression in the cochlea before the onset of hearing. Proc Natl Acad Sci U S A 97(3):1287–1292

Cordas EA, Ng L, Hernandez A, Kaneshige M, Cheng SY, Forrest D (2012) Thyroid hormone receptors control developmental maturation of the middle ear and the size of the ossicular bones. Endocrinology 153(3):1548–1560. doi:10.1210/en.2011-1834

DeLong GR, Stanbury JB, Fierro-Benitez R (1985) Neurological signs in congenital iodine-deficiency disorder (endemic cretinism). Dev Med Child Neurol 27(3):317–324

Deol MS (1973) An experimental approach to the understanding and treatment of hereditary syndromes with congenital deafness and hypothyroidism. J Med Genet 10(3):235–242

Deol MS (1976) The role of thyroxine in the differentiation of the organ of Corti. Acta Otolaryngol 81(5–6):429–435

Dettling J, Franz C, Zimmermann U, Lee SC, Bress A, Brandt N, Feil R, Pfister M, Engel J, Flamant F, Ruttiger L, Knipper M (2014) Autonomous functions of murine thyroid hormone receptor TRalpha and TRbeta in cochlear hair cells. Mol Cell Endocrinol 382(1):26–37. doi:10.1016/j.mce.2013.08.025

Dumitrescu AM, Refetoff S (2013) The syndromes of reduced sensitivity to thyroid hormone. Biochim Biophys Acta 1830(7):3987–4003. doi:10.1016/j.bbagen.2012.08.005

Ferrara AM, Onigata K, Ercan O, Woodhead H, Weiss RE, Refetoff S (2012) Homozygous thyroid hormone receptor beta-gene mutations in resistance to thyroid hormone: three new cases and review of the literature. J Clin Endocrinol Metab 97(4):1328–1336. doi:10.1210/jc.2011-2642

Forrest D, Erway LC, Ng L, Altschuler R, Curran T (1996) Thyroid hormone receptor beta is essential for development of auditory function. Nat Genet 13(3):354–357. doi:10.1038/ng0796-354

Fraser JS (1932) The pathology of deaf-mutism. Proc R Soc Med 25(6):861–878

Gauthier K, Plateroti M, Harvey CB, Williams GR, Weiss RE, Refetoff S, Willott JF, Sundin V, Roux JP, Malaval L, Hara M, Samarut J, Chassande O (2001) Genetic analysis reveals different functions for the products of the thyroid hormone receptor alpha locus. Mol Cell Biol 21(14):4748–4760. doi:10.1128/MCB.21.14.4748-4760.2001

Gillespie PG, Muller U (2009) Mechanotransduction by hair cells: models, molecules, and mechanisms. Cell 139(1):33–44. doi:10.1016/j.cell.2009.09.010

Griffith AJ, Szymko YM, Kaneshige M, Quinonez RE, Kaneshige K, Heintz KA, Mastroianni MA, Kelley MW, Cheng SY (2002) Knock-in mouse model for resistance to thyroid hormone (RTH): an RTH mutation in the thyroid hormone receptor beta gene disrupts cochlear morphogenesis. J Assoc Res Otolaryngol 3(3):279–288. doi:10.1007/s101620010092

Hinojosa R (1977) A note on development of Corti's organ. Acta Otolaryngol 84(3–4):238–251

Jones I, Srinivas M, Ng L, Forrest D (2003) The thyroid hormone receptor beta gene: structure and functions in the brain and sensory systems. Thyroid 13(11):1057–1068. doi:10.1089/105072503770867228

Kaneshige M, Kaneshige K, Zhu X, Dace A, Garrett L, Carter TA, Kazlauskaite R, Pankratz DG, Wynshaw-Boris A, Refetoff S, Weintraub B, Willingham MC, Barlow C, Cheng S (2000) Mice with a targeted mutation in the thyroid hormone beta receptor gene exhibit impaired growth and resistance to thyroid hormone. Proc Natl Acad Sci U S A 97(24):13209–13214. doi:10.1073/pnas.230285997

Kaneshige M, Suzuki H, Kaneshige K, Cheng J, Wimbrow H, Barlow C, Willingham MC, Cheng S (2001) A targeted dominant negative mutation of the thyroid hormone alpha 1 receptor causes increased mortality, infertility, and dwarfism in mice. Proc Natl Acad Sci U S A 98(26):15095–15100. doi:10.1073/pnas.261565798

Kelley MW (2006) Regulation of cell fate in the sensory epithelia of the inner ear. Nat Rev Neurosci 7(11):837–849. doi:10.1038/nrn1987

Killick R, Richardson GP (1997) Antibodies to the sulphated, high molecular mass mouse tectorin stain hair bundles and the olfactory mucus layer. Hear Res 103(1–2):131–141

Knipper M, Bandtlow C, Gestwa L, Kopschall I, Rohbock K, Wiechers B, Zenner HP, Zimmermann U (1998) Thyroid hormone affects Schwann cell and oligodendrocyte gene expression at the glial transition zone of the VIIIth nerve prior to cochlea function. Development 125(18):3709–3718

Knipper M, Gestwa L, Ten Cate WJ, Lautermann J, Brugger H, Maier H, Zimmermann U, Rohbock K, Kopschall I, Wiechers B, Zenner HP (1999) Distinct thyroid hormone-dependent expression of TrKB and p75NGFR in nonneuronal cells during the critical TH-dependent period of the cochlea. J Neurobiol 38(3):338–356

Knipper M, Richardson G, Mack A, Muller M, Goodyear R, Limberger A, Rohbock K, Kopschall I, Zenner HP, Zimmermann U (2001) Thyroid hormone-deficient period prior to the onset of hearing is associated with reduced levels of beta-tectorin protein in the tectorial membrane: implication for hearing loss. J Biol Chem 276(42):39046–39052. doi:10.1074/jbc.M103385200

Koenig RJ, Lazar MA, Hodin RA, Brent GA, Larsen PR, Chin WW, Moore DD (1989) Inhibition of thyroid hormone action by a non-hormone binding c-erbA protein generated by alternative mRNA splicing. Nature 337(6208):659–661. doi:10.1038/337659a0

Kros CJ, Ruppersberg JP, Rusch A (1998) Expression of a potassium current in inner hair cells during development of hearing in mice. Nature 394(6690):281–284. doi:10.1038/28401

Lautermann J, ten Cate WJ (1997) Postnatal expression of the alpha-thyroid hormone receptor in the rat cochlea. Hear Res 107(1–2):23–28

Lazar MA, Hodin RA, Darling DS, Chin WW (1988) Identification of a rat c-erbA alpha-related protein which binds deoxyribonucleic acid but does not bind thyroid hormone. Mol Endocrinol 2(10):893–901

Leao RN, Sun H, Svahn K, Berntson A, Youssoufian M, Paolini AG, Fyffe RE, Walmsley B (2006) Topographic organization in the auditory brainstem of juvenile mice is disrupted in congenital deafness. J Physiol 571(Pt 3):563–578. doi:10.1113/jphysiol.2005.098780

Legan PK, Rau A, Keen JN, Richardson GP (1997) The mouse tectorins. Modular matrix proteins of the inner ear homologous to components of the sperm-egg adhesion system. J Biol Chem 272(13):8791–8801

Legan PK, Lukashkina VA, Goodyear RJ, Kossi M, Russell IJ, Richardson GP (2000) A targeted deletion in alpha-tectorin reveals that the tectorial membrane is required for the gain and timing of cochlear feedback. Neuron 28(1):273–285

Leger J, Ecosse E, Roussey M, Lanoe JL, Larroque B (2011) Subtle health impairment and socio-educational attainment in young adult patients with congenital hypothyroidism diagnosed by neonatal screening: a longitudinal population-based cohort study. J Clin Endocrinol Metab 96(6):1771–1782. doi:10.1210/jc.2010-2315

Lichtenberger-Geslin L, Dos Santos S, Hassani Y, Ecosse E, Van Den Abbeele T, Leger J (2013) Factors associated with hearing impairment in patients with congenital hypothyroidism treated since the neonatal period: a national population-based study. J Clin Endocrinol Metab 98(9):3644–3652. doi:10.1210/jc.2013-1645

Lukashkin AN, Richardson GP, Russell IJ (2010) Multiple roles for the tectorial membrane in the active cochlea. Hear Res 266(1–2):26–35. doi:10.1016/j.heares.2009.10.005

Mallo M (2001) Formation of the middle ear: recent progress on the developmental and molecular mechanisms. Dev Biol 231(2):410–419. doi:10.1006/dbio.2001.0154

McKinley MP, O'Loughlin VD (2012) Human anatomy, 3rd edn. McGraw-Hill, New York

Morte B, Manzano J, Scanlan T, Vennstrom B, Bernal J (2002) Deletion of the thyroid hormone receptor alpha 1 prevents the structural alterations of the cerebellum induced by hypothyroidism. Proc Natl Acad Sci U S A 99(6):3985–3989. doi:10.1073/pnas.062413299

Mustapha M, Weil D, Chardenoux S, Elias S, El-Zir E, Beckmann JS, Loiselet J, Petit C (1999) An alpha-tectorin gene defect causes a newly identified autosomal recessive form of sensorineural pre-lingual non-syndromic deafness, DFNB21. Hum Mol Genet 8(3):409–412

Mustapha M, Fang Q, Gong TW, Dolan DF, Raphael Y, Camper SA, Duncan RK (2009) Deafness and permanently reduced potassium channel gene expression and function in hypothyroid Pit1dw mutants. J Neurosci 29(4):1212–1223. doi:10.1523/JNEUROSCI.4957-08.2009

Ng L, Rusch A, Amma LL, Nordstrom K, Erway LC, Vennstrom B, Forrest D (2001) Suppression of the deafness and thyroid dysfunction in Thrb-null mice by an independent mutation in the Thra thyroid hormone receptor alpha gene. Hum Mol Genet 10(23):2701–2708

Ng L, Goodyear RJ, Woods CA, Schneider MJ, Diamond E, Richardson GP, Kelley MW, Germain DL, Galton VA, Forrest D (2004) Hearing loss and retarded cochlear development in mice lacking type 2 iodothyronine deiodinase. Proc Natl Acad Sci U S A 101(10):3474–3479. doi:10.1073/pnas.0307402101

Ng L, Hernandez A, He W, Ren T, Srinivas M, Ma M, Galton VA, St Germain DL, Forrest D (2009) A protective role for type 3 deiodinase, a thyroid hormone-inactivating enzyme, in cochlear development and auditory function. Endocrinology 150(4):1952–1960. doi:10.1210/en.2008-1419

Ng L, Kelley MW, Forrest D (2013) Making sense with thyroid hormone—the role of T(3) in auditory development. Nat Rev Endocrinol 9(5):296–307. doi:10.1038/nrendo.2013.58

Ng L, et al. (2015). Age-related hearing loss and degeneration of cochlear hair cells in mice lacking thyroid hormone receptor beta1." Endocrinol 156(10):3853–3865.

Oppenheimer JH, Schwartz HL, Strait KA (1994) Thyroid hormone action 1994: the plot thickens. Eur J Endocrinol 130(1):15–24

Parrilla R, Mixson AJ, McPherson JA, McClaskey JH, Weintraub BD (1991) Characterization of seven novel mutations of the c-erbA beta gene in unrelated kindreds with generalized thyroid hormone resistance. Evidence for two "hot spot" regions of the ligand binding domain. J Clin Invest 88(6):2123–2130. doi:10.1172/JCI115542

Pyott SJ, Meredith AL, Fodor AA, Vazquez AE, Yamoah EN, Aldrich RW (2007) Cochlear function in mice lacking the BK channel alpha, beta1, or beta4 subunits. J Biol Chem 282(5):3312–3324. doi:10.1074/jbc.M608726200

Quignodon L, Vincent S, Winter H, Samarut J, Flamant F (2007) A point mutation in the activation function 2 domain of thyroid hormone receptor alpha1 expressed after CRE-mediated recombination partially recapitulates hypothyroidism. Mol Endocrinol 21(10):2350–2360. doi:10.1210/me.2007-0176

Raphael Y, Altschuler RA (2003) Structure and innervation of the cochlea. Brain Res Bull 60(5–6):397–422

Rau A, Legan PK, Richardson GP (1999) Tectorin mRNA expression is spatially and temporally restricted during mouse inner ear development. J Comp Neurol 405(2):271–280

Refetoff S, DeWind LT, DeGroot LJ (1967) Familial syndrome combining deaf-mutism, stuppled epiphyses, goiter and abnormally high PBI: possible target organ refractoriness to thyroid hormone. J Clin Endocrinol Metab 27(2):279–294. doi:10.1210/jcem-27-2-279

Rovet J, Walker W, Bliss B, Buchanan L, Ehrlich R (1996) Long-term sequelae of hearing impairment in congenital hypothyroidism. J Pediatr 128(6):776–783

Rubel EW, Fritzsch B (2002) Auditory system development: primary auditory neurons and their targets. Annu Rev Neurosci 25:51–101. doi:10.1146/annurev.neuro.25.112701.142849

Rusch A, Erway LC, Oliver D, Vennstrom B, Forrest D (1998) Thyroid hormone receptor beta-dependent expression of a potassium conductance in inner hair cells at the onset of hearing. Proc Natl Acad Sci U S A 95(26):15758–15762

Rusch A, Ng L, Goodyear R, Oliver D, Lisoukov I, Vennstrom B, Richardson G, Kelley MW, Forrest D (2001) Retardation of cochlear maturation and impaired hair cell function caused by deletion of all known thyroid hormone receptors. J Neurosci 21(24):9792–9800

Sakurai A, Miyamoto T, Refetoff S, DeGroot LJ (1990) Dominant negative transcriptional regulation by a mutant thyroid hormone receptor-beta in a family with generalized resistance to thyroid hormone. Mol Endocrinol 4(12):1988–1994. doi:10.1210/mend-4-12-1988

Sap J, Munoz A, Damm K, Goldberg Y, Ghysdael J, Leutz A, Beug H, Vennstrom B (1986) The c-erb-A protein is a high-affinity receptor for thyroid hormone. Nature 324(6098):635–640. doi:10.1038/324635a0

Schoenmakers N, Moran C, Peeters RP, Visser T, Gurnell M, Chatterjee K (2013) Resistance to thyroid hormone mediated by defective thyroid hormone receptor alpha. Biochim Biophys Acta 1830(7):4004–4008. doi:10.1016/j.bbagen.2013.03.018

Schwartz HL, Strait KA, Ling NC, Oppenheimer JH (1992) Quantitation of rat tissue thyroid hormone binding receptor isoforms by immunoprecipitation of nuclear triiodothyronine binding capacity. J Biol Chem 267(17):11794–11799

Sharlin DS, Visser TJ, Forrest D (2011) Developmental and cell-specific expression of thyroid hormone transporters in the mouse cochlea. Endocrinology 152(12):5053–5064. doi:10.1210/en.2011-1372

Sprenkle PM, McGee J, Bertoni JM, Walsh EJ (2001a) Consequences of hypothyroidism on auditory system function in Tshr mutant (hyt) mice. J Assoc Res Otolaryngol 2(4):312–329

Sprenkle PM, McGee J, Bertoni JM, Walsh EJ (2001b) Development of auditory brainstem responses (ABRs) in Tshr mutant mice derived from euthyroid and hypothyroid dams. J Assoc Res Otolaryngol 2(4):330–347

Szarama KB, Gavara N, Petralia RS, Chadwick RS, Kelley MW (2013) Thyroid hormone increases fibroblast growth factor receptor expression and disrupts cell mechanics in the developing organ of Corti. BMC Dev Biol 13:6. doi:10.1186/1471-213X-13-6

Tritsch NX, Bergles DE (2010) Developmental regulation of spontaneous activity in the Mammalian cochlea. J Neurosci 30(4):1539–1550. doi:10.1523/JNEUROSCI.3875-09.2010

Tritsch NX, Yi E, Gale JE, Glowatzki E, Bergles DE (2007) The origin of spontaneous activity in the developing auditory system. Nature 450(7166):50–55. doi:10.1038/nature06233

Uziel A, Rabie A, Marot M (1980) The effect of hypothyroidism on the onset of cochlear potentials in developing rats. Brain Res 182(1):172–175

Uziel A, Gabrion J, Ohresser M, Legrand C (1981) Effects of hypothyroidism on the structural development of the organ of Corti in the rat. Acta Otolaryngol 92(5–6):469–480

Uziel A, Legrand C, Ohresser M, Marot M (1983) Maturational and degenerative processes in the organ of Corti after neonatal hypothyroidism. Hear Res 11(2):203–218

van Mullem A, van Heerebeek R, Chrysis D, Visser E, Medici M, Andrikoula M, Tsatsoulis A, Peeters R, Visser TJ (2012) Clinical phenotype and mutant TRalpha1. N Engl J Med 366(15):1451–1453. doi:10.1056/NEJMc1113940

van Mullem AA, Chrysis D, Eythimiadou A, Chroni E, Tsatsoulis A, de Rijke YB, Visser WE, Visser TJ, Peeters RP (2013) Clinical phenotype of a new type of thyroid hormone resistance caused by a mutation of the TRalpha1 receptor: consequences of LT4 treatment. J Clin Endocrinol Metab 98(7):3029–3038. doi:10.1210/jc.2013-1050

Verhoeven K, Van Laer L, Kirschhofer K, Legan PK, Hughes DC, Schatteman I, Verstreken M, Van Hauwe P, Coucke P, Chen A, Smith RJ, Somers T, Offeciers FE, Van de Heyning P, Richardson GP, Wachtler F, Kimberling WJ, Willems PJ, Govaerts PJ, Van Camp G (1998) Mutations in the human alpha-tectorin gene cause autosomal dominant non-syndromic hearing impairment. Nat Genet 19(1):60–62. doi:10.1038/ng0598-60

Visser WE, Friesema EC, Jansen J, Visser TJ (2008) Thyroid hormone transport in and out of cells. Trends Endocrinol Metab 19(2):50–56. doi:10.1016/j.tem.2007.11.003

Wangemann P (2006) Supporting sensory transduction: cochlear fluid homeostasis and the endo-cochlear potential. J Physiol 576(Pt 1):11–21. doi:10.1113/jphysiol.2006.112888

Weber T, Zimmermann U, Winter H, Mack A, Kopschall I, Rohbock K, Zenner HP, Knipper M (2002) Thyroid hormone is a critical determinant for the regulation of the cochlear motor protein prestin. Proc Natl Acad Sci U S A 99(5):2901–2906. doi:10.1073/pnas.052609899

Weinberger C, Thompson CC, Ong ES, Lebo R, Gruol DJ, Evans RM (1986) The c-erb-A gene encodes a thyroid hormone receptor. Nature 324(6098):641–646. doi:10.1038/324641a0

Winter H, Braig C, Zimmermann U, Geisler HS, Franzer JT, Weber T, Ley M, Engel J, Knirsch M, Bauer K, Christ S, Walsh EJ, McGee J, Kopschall I, Rohbock K, Knipper M (2006) Thyroid hormone receptors TRalpha1 and TRbeta differentially regulate gene expression of Kcnq4 and prestin during final differentiation of outer hair cells. J Cell Sci 119(Pt 14):2975–2984. doi:10.1242/jcs.03013

Winter H, Braig C, Zimmermann U, Engel J, Rohbock K, Knipper M (2007) Thyroid hormone receptor alpha1 is a critical regulator for the expression of ion channels during final differentiation of outer hair cells. Histochem Cell Biol 128(1):65–75. doi:10.1007/s00418-007-0294-6

Winter H, Ruttiger L, Muller M, Kuhn S, Brandt N, Zimmermann U, Hirt B, Bress A, Sausbier M, Conscience A, Flamant F, Tian Y, Zuo J, Pfister M, Ruth P, Lowenheim H, Samarut J, Engel J, Knipper M (2009) Deafness in TRbeta mutants is caused by malformation of the tectorial membrane. J Neurosci 29(8):2581–2587. doi:10.1523/JNEUROSCI.3557-08.2009

Zhang J, Lazar MA (2000) The mechanism of action of thyroid hormones. Annu Rev Physiol 62:439–466. doi:10.1146/annurev.physiol.62.1.439

Zoeller RT, Rovet J (2004) Timing of thyroid hormone action in the developing brain: clinical observations and experimental findings. J Neuroendocrinol 16(10):809–818. doi:10.1111/j.1365-2826.2004.01243.x

**Part III
Thyroid Hormone Disruption and
Neurodevelopment: Human Studies**

Chapter 10
The Impact of Maternal Thyrotoxicosis and Antithyroid Drug Exposure on Fetal/Neonatal Brain Development

Ines Donangelo and Gregory A. Brent

Abstract Elevated maternal thyroid hormone levels during pregnancy, as well as antithyroid drugs, can influence fetal and neonatal brain development. Elevated thyroid hormone levels in pregnancy are associated with an increased risk of miscarriage, preterm delivery, intrauterine growth restriction, preeclampsia, and heart failure. Fetal exposure to excessive maternal thyroid hormone is associated with inhibition of the fetal hypothalamic–pituitary–thyroid axis, impaired maturation of pituitary thyrotrophs and thyroid gland development, and potential disruption of neurologic development. Early diagnosis and treatment of maternal hyperthyroidism during pregnancy with antithyroid drugs improves pregnancy outcomes, fetal growth, and neurologic development. The antithyroid drugs methimazole (MMI) and carbimazole (CAB), however, are associated with a rare embryopathy that may include aplasia cutis, choanal atresia, tracheoesophageal fistulas, omphalocele, omphalomesenteric duct anomalies, and facial abnormalities. Although the number of reported cases is small relative to the number of women taking MMI or CAB at the time of conception, similar congenital abnormalities have not been reported in association with exposure to the antithyroid drug propylthiouracil (PTU). Developmental delay has been described in a subset of cases of MMI/CAB-related embryopathy, potentially due to the perinatal hypoxia associated with bilateral choanal atresia. PTU has been associated with severe maternal hepatotoxicity. The current guidelines for treatment of hyperthyroidism in pregnancy reconcile the adverse profile of antithyroid drugs in pregnancy by recommending PTU in the first

I. Donangelo, M.D., Ph.D.
Endocrinology Division, Allegheny General Hospital, Pittsburgh, PA, USA

G.A. Brent, M.D. (✉)
Department of Medicine, 111, David Geffen School of Medicine at UCLA,
11301 Wilshire Blvd, Los Angeles, CA 90073, USA
e-mail: gbrent@mednet.ucla.edu

© Springer Science+Business Media New York 2016
N. Koibuchi, P.M. Yen (eds.), *Thyroid Hormone Disruption and Neurodevelopment*,
Contemporary Clinical Neuroscience, DOI 10.1007/978-1-4939-3737-0_10

trimester, to minimize the risk for embryopathy, and MMI in the second and third trimesters, to minimize the risk of liver failure.

Keywords Thyroid hormone • Hyperthyroidism • Antithyroid drugs • Embryopathy

10.1 Introduction

Maternal hyperthyroidism is associated with adverse pregnancy outcomes and also, rarely, with congenital malformations. In addition, excessive thyroid hormone exposure during the fetal and neonatal periods may disrupt normal neurologic and motor development. Treatment of hyperthyroidism during pregnancy with antithyroid drugs (ATD) generally improves pregnancy outcomes, although excessive treatment may result in fetal hypothyroidism and abnormal neurologic development. Additionally, methimazole (Clementi et al. 1999) is associated with an embryopathy when the fetus is exposed in the first trimester, and propylthiouracil (PTU) is associated with severe maternal hepatotoxicity. An optimal level of thyroxine in the fetus is essential for normal neural maturation, which is generally delayed with hypothyroidism and accelerated with hyperthyroidism. A transient deficiency or excess of maternal thyroid hormone during pregnancy can have deleterious consequences on brain morphology in the fetus, as well as maturation of the hypothalamic–pituitary–thyroid axis (Ahmed et al. 2008). Here we will review the effects of maternal hyperthyroidism, neonatal exposure to excess thyroid hormone, and the impact of maternal use of antithyroid drugs on fetal and neonatal brain development.

10.2 Maternal Hyperthyroidism

Maternal hyperthyroidism is associated with an increased risk of spontaneous abortion, stillbirth, preterm delivery, intrauterine growth restriction, preeclampsia, and heart failure (Patil-Sisodia and Mestman 2010). A United States cohort, the Consortium on Safe Labor (2002–2008), analyzed singleton deliveries ($n = 223,512$) and found that hyperthyroidism complicates 0.2 % of all pregnancies and is associated with increased odds of preeclampsia (OR 1.78, CI 1.08–2.94), superimposed preeclampsia (OR 3.64, CI 1.82–7.29), preterm birth (OR 1.81, CI 1.32–2.49), induction of labor (OR 1.40, CI 1.06–1.86), and Intensive Care Unit admission (OR 3.70, CI 1.16–11.80) (Mannisto et al. 2013). Pregnancy outcomes improved with control of hyperthyroidism. A retrospective analysis of 181 hyperthyroid pregnant women found an increased frequency of low birth weight (<2500 g) in women with hyperthyroidism at presentation that was not controlled during pregnancy (OR 9.2, CI 5.5–16) and a lower, but still increased, rate in pregnant women whose hyperthyroidism was controlled at the time of delivery (OR 2.4, CI 1.4–4.2). Women with a history of hyperthyroidism who were euthyroid at presentation and delivery due to treatment or spontaneous remission had similar odds for low birth weight as euthyroid control women (Millar et al. 1994). Preterm delivery (OR 16.5, CI 2.10–130) and preeclampsia (OR 4.7, CI 1.1–19.7) were also more common in women with uncontrolled hyperthyroidism during pregnancy.

Fetal exposure to excessive maternal thyroid hormone during pregnancy is associated with suppression of pituitary TSH production during the fetal and neonatal periods. It has been shown that inadequately treated maternal Graves' disease may lead to abnormalities in thyroid gland development and a pattern of congenital central hypothyroidism. The chronic inhibition of the hypothalamic–pituitary–thyroid axis by excess thyroid hormone during fetal development may inhibit the normal growth and development of the fetal thyroid (Kempers et al. 2007). In a zebrafish model, exposure to excess thyroid hormone during early embryonic development was shown to reduce TSHβ gene expression in pituitary cells and promote thyrotroph cell death (Tonyushkina et al. 2014).

Most studies of the impact of thyroid hormone excess on human fetal development and pregnancy outcomes involve women with hyperthyroidism due to autoimmune thyroid disease (Patil-Sisodia and Mestman 2010; Krassas et al. 2010). The presence of thyroid peroxidase (TPO) autoantibodies in women with Hashimoto's disease is associated with an increased miscarriage rate (Abramson and Stagnaro-Green 2001). The focus in women with Graves' disease is generally on the impact of Thyroid Stimulating Immunoglobulin (TSI) on the developing fetus, but many also have TPO antibodies. Studies of pregnancy outcomes in women with hyperthyroidism, therefore, include both the effects of excess thyroid hormone levels and potentially the effects of thyroid autoantibodies.

Evaluation of families with resistance to thyroid hormone (RTH) provides a model to study the effects of thyroid hormone excess on the fetus in mothers that do not have thyroid autoimmunity. RTH is a syndrome of reduced end-organ responsiveness to thyroid hormone, associated in most individuals with a mutation in the thyroid hormone receptor (TR) β gene. Patients with thyroid hormone resistance have elevated serum thyroid hormone levels, "inappropriately normal" TSH for the elevated serum T4 and T3, and absence of typical symptoms and clinical findings of thyroid hormone excess, although generally have tachycardia. RTH may have variable other manifestations including goiter, emotional disturbances, learning disability, short stature, delayed bone age, and recurrent ear and throat infections (Weiss et al. 2010). A normal fetus carried by a mother with RTH is exposed to high maternal hormone levels. A study evaluated families from the Azores, all with RTH due to a mutation that results in an Arginine to Glutamine substitution at codon 243 in the TRβ gene. Although fertility was not impaired, affected mothers experienced a 3- to 4-fold increase in miscarriage rate, compared with normal women (unaffected first-degree relatives or spouses of affected fathers). The difference in genotype frequency in the progeny of affected mothers (proportion ~2:1 of affected: unaffected offspring) suggests that these women tend to lose more normal than affected fetuses. This was in contrast to the progeny of affected fathers, whose spouses had similar numbers of affected and unaffected offspring. In addition, unaffected infants born to RTH-affected mothers were significantly smaller than affected infants and had suppressed TSH levels. These findings indicate that high maternal thyroid hormone levels produced a direct toxic effect to the fetus leading to increased miscarriage rates and induce catabolic state during fetal life, similar to the effects of uncontrolled hyperthyroidism (Anselmo et al. 2004). If the TRβ mutation is known in an RTH mother, the fetus can be genotyped for the same mutation and the management decision during pregnancy can be made accordingly: no treatment if fetus is affected with a TRβ mutation, and treatment to reduce maternal thyroid hormone levels when carrying an unaffected fetus with no TRβ mutation (Weiss et al. 2010).

The effects of fetal thyrotoxicosis in a euthyroid mother have been described in families with a rare germline TSH-receptor gene constitutive activating mutation. In one family (Vaidya et al. 2004), the father and their two children had a germline activating TSH-receptor (TSH-R) mutation (Ser505Asp) leading to increased thyroid hormone production, while the mother was unaffected. In the two affected children, excess thyroid hormone levels were associated with premature delivery, low birth weight, as well as motor/speech delay, gastrointestinal disorders, and advanced bone age. The diagnosis was made in the first few years of life and the children were treated with thyroidectomy. The affected father was diagnosed with thyrotoxicosis at the age of 9 years, following presentation with weight loss, nervousness, tremor, proptosis, and goiter, and he was treated with partial thyroidectomy, although thyrotoxicosis recurrence and he was treated with carbimazole and external radiation. In another reported case, a TSH-R point mutation (Ile568Thr) caused neonatal thyrotoxicosis that presented with premature delivery at 35 weeks gestation, low birth weight, increased activity level, disturbed sleep, jitteriness and exaggerated startle response, poor weight gain despite adequate formula intake, and advanced bone age (Watkins et al. 2008). His TSH was suppressed and free thyroxine levels elevated on day 9 of life, but he was only started on ATD in the sixth week of life, with normalization of free T4 and total T3 levels, although TSH remained suppressed. There was marked improvement in growth at 1 year of age, and his developmental milestones were judged to be within normal limits.

10.3   Early Life Exposure to Excess Thyroid Hormone

Neonatal thyrotoxicosis is a rare condition usually caused by transplacental passage of TSI of the IgG class, to the fetus from mothers with Graves' disease. Thyrotoxicosis is seen in about 0.1–0.2 % of pregnancies, and overt neonatal thyrotoxicosis complicates about 1–2 % of pregnancies of mothers with Graves' disease (Peters and Hindmarsh 2007). The incidence of biochemical evidence of hyperthyroidism in the offspring of mothers with Graves' disease is likely much higher, up to 20 %. Thyroid receptor antibodies may persist in Graves' patients with inactive disease after treatment with surgery, radioactive iodine, or antithyroid medication, and cross the placenta stimulating the fetal thyroid gland. Women who have had definitive radioablative therapy or surgery for Graves' disease before pregnancy tend to have more severe Graves' disease associated with larger thyroid glands and higher levels of T3, so may have especially high levels of TSI. The fetus exposed to sufficiently high levels of maternal thyroid receptor antibodies develops clinical features of excess thyroid hormone production, including goiter, tachycardia, hyperkinesis, and intrauterine growth restriction. Overt symptoms and signs usually occur within the first 10 days of life and may last for 3–6 months, until maternal IgG is cleared. Affected neonates may have irritability, restlessness, goiter, excessive weight loss, failure to regain birth weight, diarrhea, sweating, flushing, craniosynostosis, and eye signs (periorbital edema, lid retraction, and proptosis). Neonatal thyrotoxicosis is associated with a high mortality of 16–25 %, especially without early diagnosis and treatment. Cardiovascular compromise with arrhythmias, hypertension, and heart failure is the main cause of mortality (Peters and Hindmarsh 2007; Smith et al. 2001). Early diagnosis and treatment with antithyroid drugs is associated with a good prognosis. In a retrospective study, children with appropriately treated neonatal thyroxicosis had normal growth and neurologic development in the first year of life (Vautier et al. 2007).

A delay in the diagnosis and treatment of congenital hypothyroidism significantly impairs cognitive development, as discussed previously (see Chap. 13). The effects of overtreatment of congenital hypothyroidism, with excess levothyroxine early in life, have been less well studied, but also appear to be detrimental. A group of 61 children with congenital hypothyroidism in the Netherlands were assessed for the impact of levothyroxine undertreatment and overtreatment on their cognitive development scores (Bongers-Schokking et al. 2013). These children received standard treatment by their local pediatrician. They were psychologically tested at 1.8 (Mental Development Index), 6 (IQ6), and 11 years (IQ11) of age, and scores for cognitive development were correlated to the rate of initial TSH normalization, and to the total duration of undertreatment or overtreatment episodes within the first 2 years of life. Patients with a slow rate of initial TSH normalization (>2 months) had lower cognitive scores at age 1.8 years, when compared to patients with fast ($\leq$1 month) and moderate (1–2 months) rates of TSH normalization (fast and moderate TSH children had developmental scores 13.3 and 7.1 higher, respectively, than children with slow rate of TSH normalization, $p = 0.001$). However, the rate of initial TSH normalization had no effect on IQ at age 11 years. Overtreatment (>2 SD of individual steady-state concentration FT4) of congenital hypothyroidism based on serum FT4 concentration for long (>3 months) or short ($\leq$3 months) periods of time had no effect on developmental scores at 1.8 and 6 years, but was associated with lower IQ scores at 11 years (−17.8 and −13.4 points lower in long and short overtreated patients, respectively, compared to patients treated at a usual replacement dose, $p = 0.014$). This suggests that steady-state serum-free T4 concentration may be a more sensitive indicator of adverse outcome of overtreatment with thyroid hormone than TSH levels. Interestingly, short or long periods of undertreatment during the first 2 years of life did not affect developmental scores, but undertreatment with periods of overtreatment was associated with −14.7 points lower IQ scores at age 11 years.

A Swiss study evaluated the intellectual outcome, at age 14 years, of 63 children with congenital hypothyroidism treated with early high-dose thyroxine (median 14.7 μg/Kg/day, range 9.9–23.6), comparing control children with similar age and from the same geographical area (Dimitropoulos et al. 2009). They noted that despite early high initial treatment and optimal levothyroxine substitution during childhood, a significant proportion of children with congenital hypothyroidism had intellectual deficits in adolescence, when compared to their peers (adjusted IQ score 101.7 vs. 111.4, in congenital hypothyroidism vs. controls, respectively, $p<0.05$). This developmental gap was more pronounced in children with complete absence of the thyroid and lower socioeconomic status. This may be due, in part, to thyroid hormone deficit during the prenatal brain development. In contrast to the previous study, there was no correlation between under- or overtreatment with thyroid hormone replacement in the first years of life, and IQ at 14 years of age.

10.4 Maternal Use of Antithyroid Drugs

Untreated hyperthyroidism in pregnant women is associated with serious risks for both fetus and mother that are generally reduced by control of maternal thyroid hormone levels (Millar et al. 1994). Treatment options for hyperthyroidism include antithyroid medication, surgical resection of the thyroid gland, and radioactive iodine treatment. Antithyroid drugs are considered the treatment modality of choice during pregnancy, given that radioactive iodine treatment is contraindicated during pregnancy and the surgical approach is best during the second trimester and usually considered a second-line option in patients that fail or cannot take ATD (De Groot et al. 2012; Stagnaro-Green et al. 2011). The two ATDs that have been used during pregnancy are propylthiouracil (PTU) and methimazole (Clementi et al. 1999). Both drugs readily cross the placenta with similar kinetics of placental transfer (Mortimer et al. 1997). Carbimazole, used extensively in Europe, is a pro-drug and after absorption is converted to the active form, methimazole.

The effects of these drugs on human embryological and fetal development have been investigated in cohort studies, case–control studies, and case reports. The retrospective nature of these studies limits clear differentiation between drug-specific effects and those potentially related to other factors with hyperthyroidism.

10.4.1 Methimazole and Carbimazole

A systematic review on the effects of MMI and Carbimazone (CMZ) during pregnancy (Hackmon et al. 2012) identified several retrospective cohort studies, some containing hundreds of patients, evaluating the frequency of congenital malformations after intrauterine exposure of these ATDs. The authors concluded that because of the rare occurrence of aplasia cutis congenital (heterogenous conditions with loss of cutaneous structures in the crown of the head) and choanal atresia (blocking or narrowing of nasal passages), cohort studies could not detect a clear association between these malformation and prenatal exposure to MMI. In contrast, a case–control study found a significant association between MMI exposure and choanal atresia, but not with aplasia cutis congenita (Barbero et al. 2008). This multicenter study compared the frequency of maternal hyperthyroidism treated with MMI during pregnancy in children with choanal atresia compared with control group, and found that treatment with MMI was identified in 16 % of the cases (10/61) and only 1 % (2/183) in the control group (OR 17.75, 95 % CI 3.49–121.4). There was no difference between cases and control group with respect to parental degree of education, paternal occupation, twinning, maternal parity, and other exposures during pregnancy. The authors conclude that prenatal exposure to treatment with MMI is associated with choanal atresia. In some cases, exposure to MMI occurred late in pregnancy rather than during the early critical embryogenic period. In these cases, it is likely that the mother's hyperthyroidism, and not MMI treatment, is the causal factor (Barbero et al. 2008). Another case–control study reviewed 49,091 birth records of neonates born in Amsterdam and reported 13 cases of scalp skin defects (0.03 %) (Van Dijke et al. 1987). Examination of patient files showed that none of the mothers of these children had used antithyroid drugs. In addition, records of children of 24 mothers who had received MMI or carbimazole treatment in the first trimester of pregnancy showed no signs of skin defects, suggesting that there is no clear association between these drugs and congenital skin defects (Van Dijke et al. 1987).

Recent population studies add to the evidence that MMI/CMZ exposure in early pregnancy might be associated with a specific embryopathy (Andersen et al. 2013; Yoshihara et al. 2012). A Danish population-based cohort study evaluated the outcome of live-born children between 1996 and 2008 with respect to maternal exposure to ATDs during pregnancy. Of the 817,093 children included in the study, 0.22 % were exposed to maternal ATD in early pregnancy (PTU, $n = 564$; MMI/CMZ, $n = 1097$; MMI/CMZ and PTU, $n = 159$) (Andersen et al. 2013). The overall prevalence of birth defects was significantly higher in children exposed to ATD early in pregnancy compared to those not exposed (PTU, 8.0 %; MMI/CMZ, 9.1 %; MMI/CMZ and PTU, 10.1 %; no ATD during pregnancy, 5.4 %; nonexposed, 5.7 %; $p < 0.001$). Choanal atresia, esophageal atresia, omphalocele, omphalomesenteric duct anomalies, and aplasia cutis were common in MMI/CMZ-exposed children (combined, adjusted OR = 21.8 [13.4–35.4]) with 1.60 % of the MMI/CMZ-exposed children (20 of 1256) developing these malformations. Of note, in the small group of women who changed from MMI/CMZ to PTU, during their pregnancy, there was no amelioration of birth defects. There was no difference in nervous system birth defects among ADT-exposed and nonexposed children. Although maternal thyroid hormone levels were not available in this study, the authors proposed that the difference in pattern of birth defects between MMI/CMZ and PTU suggests that the birth defects are caused by ATD treatment and not by abnormal thyroid function. A Japanese study evaluated outcomes of 6744 women with Graves' disease who became pregnant with resulting 5967 live births (Yoshihara et al. 2012). They found that the overall rate of major congenital anomalies was higher in patients treated with MMI, when compared to patients in remission or previously treated [4.1 % MMI vs. 2.1 % control, OR 2.28 (1.54–3.33)]. There was no increase in the overall rate of major anomalies in the PTU group in comparison with the control group, as discussed later. Seven of the 1231 newborns in the MMI group had aplasia cutis congenita, six had an omphalocele, seven had a symptomatic omphalomesenteric duct anomaly, and one had esophageal atresia. First trimester FT4 levels were higher in MMI group (FT4 1.41 ± 0.91 and 1.29 ± 0.41 ng/dL in MMI vs. control, respectively, $p < 0.000$); however, hyperthyroidism in the first trimester of pregnancy did not increase the rate of congenital malformation.

In summary, MMI and CMZ have been associated with aplasia cutis congenita, choanal atresia, tracheoesophageal fistulas, and other less common anomalies, including facial abnormalities (upslanting palpebral fissures with epicanthic folds, arched flared eyebrows, and small nose with broad nasal bridge), hypoplastic nipples, esophageal atresia, growth restriction, circulatory or urinary system malformations, omphalocele or omphalomesenteric duct anomalies, delayed development (Hackmon et al. 2012; Andersen et al. 2013; Yoshihara et al. 2012). The medical literature describes several cases of children or fetuses exposed to MMI or CMZ in the first trimester of pregnancy, with some or all of the previously mentioned malformations suggesting a drug-specific embryopathy (Clementi et al. 1999; Barbero et al. 2004; Barwell et al. 2002; Foulds et al. 2005; Greenberg 1987; Johnsson et al. 1997; Ozgen et al. 2006; Valdez et al. 2007; Wilson et al. 1998; Karlsson et al. 2002; Martin-Denavit et al. 2000; Mujtaba and Burrow 1975; Ramirez et al. 1992). Although the number of reported cases is small relative to the number of women taking MMI/CMZ at the time of conception, the finding is significant in that similar congenital abnormalities have not been reported in association with PTU exposure (Mandel and Cooper 2001). Despite the association, the majority of children exposed to MMI during fetal life do not have obvious abnormalities, suggesting that additional factors, such as genetic susceptibility, may be required for the embryopathy.

10.4.2 Propylthiouracil

In contrast to observation of rare yet characteristic teratogenic effect for MMI/CMZ as described earlier, there is no definite association between PTU use during pregnancy and fetal abnormalities. A Danish population-based cohort study found a significant increased odd ratio of face, neck, and urinary system malformations in children exposed to PTU early in pregnancy [0.53 % (3/564) of face and neck and 0.89 % (5/564) of urinary system malformations in PTU vs. 0.08 % (625/811,730) and 0.30 % (2431/811,730), respectively, in nonexposed] (Andersen et al. 2013). However, other studies failed to show an association between maternal PTU use and congenital malformations. In a Japanese study that evaluated pregnancy outcome of patients with Graves' disease, there was a similar rate of congenital malformation in maternal PTU use in comparison to patients in remission or previously treated (1.9 %, 21/1399 in PTU vs. 2.1 %, 40/1906 in control, $p = 0.709$) (Yoshihara et al. 2012). A systematic review of retrospective studies on PTU exposure during pregnancy

(Hackmon et al. 2012) describes rates of major birth defects comparable to general population. One prospective observational controlled cohort study of women counseled by the Israeli Teratology Information Service between 1994 and 2004 (Rosenfeld et al. 2009) compared the rate of major anomalies between 115 PTU-exposed pregnancies and 1141 controls exposed to nonteratogens. The rate of major anomalies was comparable between the PTU-exposed (1.3 %, 1/80) and control groups (3.2 %, 34/1066). There was no difference in pregnancy outcomes with respect to miscarriage (PTU 7.8 % vs. control 6.3 %, $p=0.528$) or preterm delivery (PTU 10 % and control 5.7 %, p 0.086) rates. Median gestational age [PTU 40 weeks (38–40) vs. control 40 weeks (39–41), $p=0.018$], and median birth weight [PTU 3145 g (2655–3537) vs. control 3300 g (2968–3600), $p=0.018$] were lower. Part of the differences in neonatal birth weight between the groups could be attributed to earlier gestational age at delivery, higher rate of multiple pregnancies, female gender of the newborn, smoking and younger maternal age, but the study lacked data on maternal thyroid status during pregnancy as an important influence on fetal weight gain.

Although there is a paucity of data indicating an association between maternal PTU use and congenital malformations, PTU therapy can cause rare but severe liver toxicity, likely due to immune allergic response specific to this drug compound (Kim et al. 2001). The estimated occurrence of PTU-related acute liver failure is 1 in 10,000 exposed adults, with high mortality and need for liver transplant in severe cases. In 2010, the FDA issued a "black box" warning, due to hepatotoxicity, recommending that PTU should not be prescribed as first-line treatment, except in the first trimester of pregnancy (due to reports on MMI teratogenicity), thyroid storm or life-threatening thyrotoxicosis (superior inhibition of peripheral conversion to T4 to T3) and when MMI, radioiodine, or surgery and not treatment options (De Groot et al. 2012; Stagnaro-Green et al. 2011). Severe PTU-related hepatotoxicity has been reported in pregnant women with Graves' disease (Morris et al. 1989; Parker 1982).

10.4.3   Antithyroid Drugs and Neurodevelopment

Eisentstein et al. (1992) evaluated the intellectual capacity of subjects born to women with Graves' disease who received antithyroid drugs throughout pregnancy with the Wechsler Intelligence test appropriate for age, compared to 25 unexposed siblings. Subjects exposed to MMI (40–140 mg/week, $n=15$) and PTU (250–1400 mg/week, $n=16$) did not differ from the unexposed group with respect to the total and subcategories of IQ, or in performance. There was also no difference between exposure to high antithyroid doses (MMI > 40 mg/week and PTU > 600 mg/week) and lower doses, and all children were euthyroid at birth and did not have a goiter. The findings in this study suggest that exposure to MMI or PTU during pregnancy in doses sufficient to control maternal hyperthyroidism does not pose a threat to intellectual capacity of offspring.

In a subset of the cases with the proposed MMI-related embryopathy, there was developmental delay (Table 10.1). Psychomotor delay in patients prenatally exposed

Table 10.1 Neurodevelopment delay in children exposed to methimazole or carbimazole in pregnancy

	Greenberg (1987)	Wilson et al. (1998)	Clementi et al. (1999)	Foulds et al. (2005)	Ozgen et al. (2006)	Valdez et al. (2007)	Gripp et al. (2011)
Mother's thyroid status during pregnancy	?	?	Euthyroid	Euthyroid	?	Hyperthyroid early pregnancy, then euthyroid	?
Treatment	MMI	Carbimazole (10–40 mg/day)	MMI (20 mg/day) until 6th GW, PTU 7th GW to birth	Carbimazole (5–20 mg/day)	MMI	MMI (50 mg/day until GW 5.5, then 30 mg/day 9th GW till birth)	MMI (15 m/day) until 6th GW / PTU (50 mg/day) 10th–16th GW / Thyroidectomy 16th GW
Gestational age (weeks)	36	34	31	36	?	40	35
Birth weight (g)	2190 (2nd–9th centile)	2070 (25–50th centile)	1470 (25–50th centile)	2900 (75th centile)	?	2270 (<3rd centile)	2750 (>90th centile)
Thyroid function	Euthyroid (3.5 years)	Not reported	Euthyroid		?	Not reported	?
Perinatal respiratory distress	?		Yes		?	No	Yes
Development	Impaired gross motor skills / IQ 68	Mild global delay, mainly motor	Severe global delay / Periventricular Leukomalacia	Mild global delay	Speech and language development delay	Neurodevelopmental delay / IQ 68	Moderate global delay
Choanal atresia	Yes	Yes	Yes	Yes	Yes	No	No

Esophageal atresia with T-E fistula	No	No	Yes	No	No	No	Yes
Facial abnormalities	Yes	Yes	Yes	Yes	Yes	Yes	Yes
Nipple development	Athelia	Hypothelia	Normal	Small but not hypoplastic	?	Normal	Normal
Other	Sensorial hearing loss Area of alopecia of parietal region		Bilateral strabismus Depigmentation of fundi	Short fifth fingers Single palmar creases Patent vitellointestinal Duct Moderate mixed hearing loss		Bilateral radioulnar synostosis	Scalp cutis aplasia Microtia and overfolded upper ear helices Clinodactyly fifth finger Right sensorineural hearing loss Possible ectopic pituitary gland

MMI methimazole, *GW* gestational week

to MMI could occur by direct action of the drug, due to abnormal maternal and fetal thyroid hormone levels early in pregnancy, or related with perinatal hypoxia as reported by Clementi et al. (1999) where the impossibility of nasal intubation led to diagnosis of bilateral choanal atresia. Notably, all (Clementi et al. 1999; Foulds et al. 2005; Greenberg 1987; Ozgen et al. 2006; Wilson et al. 1998) except one (Valdez et al. 2007) of the reported cases of developmental delay in association with MMI exposure were in infants with choanal atresia (Table 10.1), indicating that a hypoxic episode is a potential contributing factor to the neurological disabilities.

Fetal hypothyroidism and goiter may occur as complication of antithyroid drug therapy in pregnancy, even when the maternal-free T4 is in the normal range. This complication, however, can be lessened by using the lowest dose of ATD possible, aiming for a maternal-free T4 concentration in the upper end of the reference range for nonpregnant women (Momotani et al. 1986). Fortunately, children exposed to ATD in utero have not been shown to have altered growth or intellectual development. In one study, comparing 17 children of hyperthyroid mothers with 25 children of euthyroid mothers not on ATD treatment, the mean birth weight was lower in the infants of hyperthyroid mothers (Messer et al. 1990). The individual birth weights, however, were normal for gestational age and the body weight differences did not persist during development. The psychomotoric and intellectual capacity in the children of hyperthyroid mothers were normal. Similar findings were seen in another study of the intellectual capacity of 31 subjects aged 4–23 years born to women with Graves' disease who received ATD during pregnancy, as assessed using the Wechsler Intelligence test appropriate for age, was similar to that of 25 unexposed siblings (Eisenstein et al. 1992).

10.5 Summary

Both maternal hyperthyroxinemia and ATD use during pregnancy can influence fetal neurologic development and brain function. Studies in pregnant women with RTH and carrying an unaffected fetus have given an indication of the upper limit of maternal thyroxine levels associated with adverse fetal outcomes, and pregnant women treated with ATDs have identified the lower limits of maternal thyroxine levels associated with adverse fetal outcomes. Rare adverse outcomes associated with ATD use in pregnancy include MMI-associated embryopathy and PTU-associated maternal hepatic toxicity. The current guidelines for treatment of hyperthyroidism in pregnancy recommend the use of PTU in the first trimester and MMI in the second and third trimesters (De Groot et al. 2012; Stagnaro-Green et al. 2011). Although some have raised concerns about direct adverse effects of ATDs on neural development, even if thyroxine levels are normal, these effects have not been consistently identified in systematic studies. Fetal exposure to the lowest amount of ATD required for maternal treatment of hyperthyroidism, however, remains the recommended approach.

Acknowledgements   This work was supported by Veterans Affairs Merit Review Funds to GAB.

References

Abramson J, Stagnaro-Green A (2001) Thyroid antibodies and fetal loss: an evolving story. Thyroid 11:57–63

Ahmed OM, El-Gareib AW, El-Bakry AM et al (2008) Thyroid hormones states and brain development interactions. Int J Dev Neurosci 26:147–209

Andersen SL, Olsen J, Wu CS et al (2013) Birth defects after early pregnancy use of antithyroid drugs: a Danish nationwide study. J Clin Endocrinol Metab 98:4373–4381

Anselmo J, Cao D, Karrison T et al (2004) Fetal loss associated with excess thyroid hormone exposure. JAMA 292:691–695

Barbero P, Ricagni C, Mercado G et al (2004) Choanal atresia associated with prenatal methimazole exposure: three new patients. Am J Med Genet A 129A:83–86

Barbero P, Valdez R, Rodriguez H et al (2008) Choanal atresia associated with maternal hyperthyroidism treated with methimazole: a case-control study. Am J Med Genet A 146A:2390–2395

Barwell J, Fox GF, Round J et al (2002) Choanal atresia: the result of maternal thyrotoxicosis or fetal carbimazole? Am J Med Genet 111:55–56, discussion 54

Bongers-Schokking JJ, Resing WC, de Rijke YB et al (2013) Cognitive development in congenital hypothyroidism: is overtreatment a greater threat than undertreatment? J Clin Endocrinol Metab 98:4499–4506

Clementi M, Di Gianantonio E, Pelo E et al (1999) Methimazole embryopathy: delineation of the phenotype. Am J Med Genet 83:43–46

De Groot L, Abalovich M, Alexander EK et al (2012) Management of thyroid dysfunction during pregnancy and postpartum: an Endocrine Society clinical practice guideline. J Clin Endocrinol Metab 97:2543–2565

Dimitropoulos A, Molinari L, Etter K et al (2009) Children with congenital hypothyroidism: long-term intellectual outcome after early high-dose treatment. Pediatr Res 65:242–248

Eisenstein Z, Weiss M, Katz Y et al (1992) Intellectual capacity of subjects exposed to methimazole or propylthiouracil in utero. Eur J Pediatr 151:558–559

Foulds N, Walpole I, Elmslie F et al (2005) Carbimazole embryopathy: an emerging phenotype. Am J Med Genet A 132A:130–135

Greenberg F (1987) Choanal atresia and athelia: methimazole teratogenicity or a new syndrome? Am J Med Genet 28:931–934

Gripp KW, Kuryan R, Schnur RE et al (2011) Grade 1 microtia, wide anterior fontanel and novel type tracheo-esophageal fistula in methimazole embryopathy. Am J Med Genet A 155A:526–533

Hackmon R, Blichowski M, Koren G (2012) The safety of methimazole and propylthiouracil in pregnancy: a systematic review. J Obstet Gynaecol Can 34:1077–1086

Johnsson E, Larsson G, Ljunggren M (1997) Severe malformations in infant born to hyperthyroid woman on methimazole. Lancet 350:1520

Karlsson FA, Axelsson O, Melhus H (2002) Severe embryopathy and exposure to methimazole in early pregnancy. J Clin Endocrinol Metab 87:947–949

Kempers MJ, van Trotsenburg AS, van Rijn RR et al (2007) Loss of integrity of thyroid morphology and function in children born to mothers with inadequately treated Graves' disease. J Clin Endocrinol Metab 92:2984–2991

Kim HJ, Kim BH, Han YS et al (2001) The incidence and clinical characteristics of symptomatic propylthiouracil-induced hepatic injury in patients with hyperthyroidism: a single-center retrospective study. Am J Gastroenterol 96:165–169

Krassas GE, Poppe K, Glinoer D (2010) Thyroid function and human reproductive health. Endocr Rev 31:702–755

Mandel SJ, Cooper DS (2001) The use of antithyroid drugs in pregnancy and lactation. J Clin Endocrinol Metab 86:2354–2359

Mannisto T, Mendola P, Grewal J et al (2013) Thyroid diseases and adverse pregnancy outcomes in a contemporary US cohort. J Clin Endocrinol Metab 98:2725–2733

Martin-Denavit T, Edery P, Plauchu H et al (2000) Ectodermal abnormalities associated with methimazole intrauterine exposure. Am J Med Genet 94:338–340

Messer PM, Hauffa BP, Olbricht T et al (1990) Antithyroid drug treatment of Graves' disease in pregnancy: long-term effects on somatic growth, intellectual development and thyroid function of the offspring. Acta Endocrinol (Copenh) 123:311–316

Millar LK, Wing DA, Leung AS et al (1994) Low birth weight and preeclampsia in pregnancies complicated by hyperthyroidism. Obstet Gynecol 84:946–949

Momotani N, Noh J, Oyanagi H et al (1986) Antithyroid drug therapy for Graves' disease during pregnancy. Optimal regimen for fetal thyroid status. N Engl J Med 315:24–28

Morris CV, Goldstein RM, Cofer JB et al (1989) An unusual presentation of fulminant hepatic failure secondary to propylthiouracil therapy. Clin Transpl 311

Mortimer RH, Cannell GR, Addison RS et al (1997) Methimazole and propylthiouracil equally cross the perfused human term placental lobule. J Clin Endocrinol Metab 82:3099–3102

Mujtaba Q, Burrow GN (1975) Treatment of hyperthyroidism in pregnancy with propylthiouracil and methimazole. Obstet Gynecol 46:282–286

Ozgen HM, Reuvers-Lodewijks WE, Hennekam RC (2006) [Possible teratogenic effects of thiamazole]. Ned Tijdschr Geneeskd 150:101–104

Parker WA (1982) Propylthiouracil-induced hepatotoxicity. Clin Pharm 1:471–474

Patil-Sisodia K, Mestman JH (2010) Graves hyperthyroidism and pregnancy: a clinical update. Endocr Pract 16:118–129

Peters CJ, Hindmarsh PC (2007) Management of neonatal endocrinopathies—best practice guidelines. Early Hum Dev 83:553–561

Ramirez A, Espinosa de los Monteros A, Parra A et al (1992) Esophageal atresia and tracheo-esophageal fistula in two infants born to hyperthyroid women receiving methimazole (Tapazole) during pregnancy. Am J Med Genet 44:200–202

Rosenfeld H, Ornoy A, Shechtman S et al (2009) Pregnancy outcome, thyroid dysfunction and fetal goitre after in utero exposure to propylthiouracil: a controlled cohort study. Br J Clin Pharmacol 68:609–617

Smith C, Thomsett M, Choong C et al (2001) Congenital thyrotoxicosis in premature infants. Clin Endocrinol (Oxf) 54:371–376

Stagnaro-Green A, Abalovich M, Alexander E et al (2011) Guidelines of the American Thyroid Association for the diagnosis and management of thyroid disease during pregnancy and postpartum. Thyroid 21:1081–1125

Tonyushkina KN, Shen MC, Ortiz-Toro T et al (2014) Embryonic exposure to excess thyroid hormone causes thyrotrope cell death. J Clin Invest 124:321–327

Vaidya B, Campbell V, Tripp JH et al (2004) Premature birth and low birth weight associated with nonautoimmune hyperthyroidism due to an activating thyrotropin receptor gene mutation. Clin Endocrinol (Oxf) 60:711–718

Valdez RM, Barbero PM, Liascovich RC et al (2007) Methimazole embryopathy: a contribution to defining the phenotype. Reprod Toxicol 23:253–255

Van Dijke CP, Heydendael RJ, De Kleine MJ (1987) Methimazole, carbimazole, and congenital skin defects. Ann Intern Med 106:60–61

Vautier V, Moulin P, Guerin B et al (2007) [Transient neonatal thyrotoxicosis: clinical presentation and treatment in 7 cases]. Arch Pediatr 14:1310–1314

Watkins MG, Dejkhamron P, Huo J et al (2008) Persistent neonatal thyrotoxicosis in a neonate secondary to a rare thyroid-stimulating hormone receptor activating mutation: case report and literature review. Endocr Pract 14:479–483

Weiss RE, Dumitrescu A, Refetoff S (2010) Approach to the patient with resistance to thyroid hormone and pregnancy. J Clin Endocrinol Metab 95:3094–3102

Wilson LC, Kerr BA, Wilkinson R et al (1998) Choanal atresia and hypothelia following methimazole exposure in utero: a second report. Am J Med Genet 75:220–222

Yoshihara A, Noh J, Yamaguchi T et al (2012) Treatment of Graves' disease with antithyroid drugs in the first trimester of pregnancy and the prevalence of congenital malformation. J Clin Endocrinol Metab 97:2396–2403

Chapter 11
Deficit in Thyroid Hormone Transporters and Brain Development

Takehiro Suzuki and Takaaki Abe

Abstract
1. Thyroid hormone is essential for the development of various tissues, especially in the developing brain.
2. Several classes of transporters, such as organic anion transporting polypeptides (OATP1C1/Oatp1c1), monocarboxylate transporters (Mct8/MCT8), and amino acid transporters (Lat1/LAT1), are expressed in different components of the central nervous system (i.e., blood brain barrier, astrocytes, and neurons), functioning concertedly (see Fig. 11.1).
3. The fact that only MCT8 and Oatp1c1 double knockout mice displayed more severely decreased uptakes and contents of T3 and T4 in brain as well as exhibited more impaired brain development phenotypes (delayed cerebellar development, reduced myelination, and abnormal locomotor activities), suggesting compensations and redundancies of thyroid hormone transportation in the brain by several different transporters.

Keywords Thyroid hormone transporter • Organic anion transporting polypeptide • Monocarboxylate transporter • Amino acid transporter • Type2 deiodinase • Type3 deiodinase • Blood–brain barrier

T. Suzuki
Division of Nephrology, Endocrinology, and Vascular Medicine, Department of Medicine, Tohoku University Graduate School of Medicine, Sendai, Miyagi, Japan

Department of Medical Science, Tohoku University Graduate School of Biomedical Engineering, Sendai, Miyagi, Japan
e-mail: suzuki2i@med.tohoku.ac.jp

T. Abe (✉)
Division of Nephrology, Endocrinology, and Vascular Medicine, Department of Medicine, Tohoku University Graduate School of Medicine, Sendai, Miyagi, Japan

Division of Medical Science, Tohoku University Graduate School of Biomedical Engineering, 1-1 Siryo-cho, Aoba-ku, Sendai, Miyagi 980-8574, Japan

Department of Clinical Biology and Hormonal Regulation, Tohoku University Graduate School of Medicine, Sendai, Miyagi, Japan
e-mail: takaabe@med.tohoku.ac.jp

© Springer Science+Business Media New York 2016
N. Koibuchi, P.M. Yen (eds.), *Thyroid Hormone Disruption and Neurodevelopment*, Contemporary Clinical Neuroscience, DOI 10.1007/978-1-4939-3737-0_11

Abbreviations

BBB	Blood–brain barrier
CSF	Cerebrospinal fluid
MCT	Monocarboxylate transporter
ntcp	Na^+/taurocholate cotransporting polypeptide
oatp	Organic anion transporting polypeptide
T4, thyroxine	T3, 3,3′, 5-triiodo-L-thyronine

11.1 Introduction

Thyroid hormone is important for the development of various tissues and for the regulation of metabolic processes throughout life (Porterfield and Hendrich 1993; Oppenheimer and Schwartz 1997; Yen 2001). In the developing brain, it was reported that thyroid hormone regulates about 500–1000 genes (Chatonnet et al. 2015) and the process of terminal brain differentiation such as axonal and dendritic growth, synaptogenesis, neural migration, and myelination (Eayrs 1960). Therefore, thyroid hormone deficiency in the developmental period results in severe functional defects of the brain, including mental retardation, ataxia, spasticity, and deafness (Thompson and Potter 2000). In thyroid hormone-deficient rat, cells in the cortex are smaller and more closely aggregated than normal, due in part to an overall decrease in the development of axonal and dendritic processes (Eayrs 1960). Axonal density is also decreased and the probability of axo-dendritic interaction is reduced (Eayrs 1960). In the cerebellum, both cell migration and differentiation of granule cells are affected (Oppenheimer and Schwartz 1997). A deficiency of myelination under hypothyroid status has been observed in the cerebral cortex, visual and auditory cortex, hippocampus, and cerebellum, which were related to the observed neurodevelopmental delay (Balazs et al. 1969, 1971).

The brain is separated from the bloodstream by the blood–brain and blood–cerebrospinal fluid (CSF) barriers, which restrict the entry of molecules into brain tissue. It has been suggested that thyroid hormone enters the brain via the blood–brain barrier (BBB) or the blood–CSF barrier (Southwell et al. 1993). The choroid plexus is a major component of the blood–CSF barrier, as well as the major production site of the thyroid hormone-binding protein, transthyretin. Free T4 can cross the choroid plexus from blood to the brain (Southwell et al. 1993) and one-fifth of the T4 in the brain could originate from CSF that has passed through the choroid plexus (Chanoine et al. 1992). In the retina, the retinal pigment epithelium is the only source of transthyretin and is supposed to be a functional counterpart of choroid plexus epithelium in the blood retinal barrier (Cavallaro et al. 1990). However, the molecular mechanisms of thyroid hormone transport in the BBB and blood–CSF barriers have not been clarified. Although the thyroid hormone, being hydrophobic, had been thought to enter target cell membranes by passive diffusion, the evidence has accumulated over the past three decades for the existence of multiple thyroid hormone transporting systems in different tissues, such as liver (Krenning et al. 1981; Blondeau et al. 1988; Topliss et al. 1989), neuronal cells (Chantoux et al. 1995), pituitary (Gingrich et al. 1985), astrocytes (Beslin et al. 1995), glial cells (Francon et al. 1989), skeletal muscles (Centanni and Robbins 1987), red blood cells (Galton et al. 1986), and erythrocytes (Osty et al. 1988) with or without Na^+ dependency, and the Km values varies from the nM to µM level. Because both the action and metabolism of thyroid hormone are intracellular events, an uptake system of thyroid hormone through the plasma membrane should be required (Abe et al. 2002; Friesema et al. 2005; Hennemann et al. 2001). On the contrary, transthyretin has been recognized as a major thyroid hormone-binding protein in the CSF in human and rodents, its importance as a carrier for transporting T4 into the brain was recently questioned since transthyretin-deficient mice exhibit normal T3 and T4 concentrations in the brain parenchyma (Palha et al. 2000). These data suggest that thyroid hormone-binding proteins do not represent a limiting factor for thyroid hormone to be normally distributed to the central nervous system.

11.2   Oatp1a4 and Oatp1a5 as Thyroid Hormone Transporters

The characterization of thyroid hormone transport was started because of the structural and functional similarities between the choroid plexus epithelium and the retinal pigment epithelium. In 1998, Abe et al. isolated Oatp1a4(oatp2) and Oatp1a5(oatp3) from rat retina and identified them as thyroid hormone transporters (Abe et al. 1998). Hydrophobicity analyses of Oatp1a4 and Oatp1a5 predict 12 hydrophobic segments, transport T4 and T3. The Km values for T4 and T3 by Oatp1a4 were 6.5 µM and 5.8 µM, respectively. Oatp1a5 also transports thyroid hormone with a Km of 4.9 µM for T4 and 7.3 µM for T3. Neither Oatp1a4- nor Oatp1a5-mediated thyroid hormone uptake is dependent on extracellular Na^+. In addition, immunohistochemical analysis showed that Oatp1a4 is localized mainly in the retinal pigment epithelium, whereas Oatp1a5 is located in optic nerve fibers, suggesting that they might play successive roles in thyroid hormone transport in the retina and neural cells (Ito et al. 2002).

11.3   Thyroid Hormone Transporting OATP Family in BBB and Blood–CSF Barrier

Subsequently, rat Oatp1a1(oatp1), which was originally identified as the bile acid transporter in the rat liver (Jacquemin et al. 1994) also reported to transport thyroid hormone in a Na^+-independent manner (Friesema et al. 1999). Localization of Oatp1a1 and Oatp1a4 at the apical (Angeletti et al. 1997) and basolateral membranes

(Gao et al. 1999) of the choroid plexus epithelium was reported, respectively. This spatial localization suggests the role for the Oatps in the transport of thyroid hormone between the blood, CSF, and brain. Recently, Oatp1a5 expressions in rat (Ohtsuki et al. 2003; Kusuhara et al. 2003) and in mouse (Ohtsuki et al. 2004) apical membrane of choroid plexus epithelial cells were also reported. In these reports, as the Oatp1a1 expression in choroid plexus was reported to be very low or under detectable level in mRNA level (Ohtsuki et al. 2003, 2004; Kusuhara et al. 2003), so the expression of Oatp1a1 in choroid plexus epithelial and the involvement in blood–CSF barrier transportation of thyroid hormone remain to be examined. Oatp1a5 might be another candidate for thyroid hormone transporting system in the apical side of choroid plexus epithelial cells.

In human, it was also reported that OATP1C1(OATP-F), which is highly expressed in the brain and testis, transports T4 ($Km=90$ nM) and reverse T3 ($Km=128$ nM) in a high affinity manner (Pizzagalli et al. 2002). In the brain, OATP-1C1 mRNA was detected in numerous brain regions except the pons and cerebellum (Pizzagalli et al. 2002). Immunohistochemical study revealed that OATP1C1 was also localized in Leydig cells of human testis, although immunohistochemical study in human brain was not performed in this report (Pizzagalli et al. 2002). The extensive expression of OATP1C1 expressed throughout the brain, except for the cerebellum and pons as well as high affinity for T4, suggests that this transporter should be involved in the brain uptake of T4.

The rat and mouse ortholog of human OATP1C1(OATP-F), Oatp1c1(oatp14) is localized preferentially in brain capillary endothelial cells capillaries in rat (Sugiyama et al. 2003) and in mouse (Tohyama et al. 2004). Increased transport was demonstrated for estradiol 17-β-glucuronide, cerivastatin, troglitazone sulfate, T4 and rT3 in Oatp1c1 transfected HEK293 cells. Km and Vmax values for these substrates show the highest affinity and specificity for T4 and rT3. Sugiyama and colleagues also reported the altered expression level of Oatp1c1 in isolated brain capillaries in hypothyroid and hyperthyroid rat models (Sugiyama et al. 2003). In hypothyroid rats, Oatp1c1 mRNA and protein are upregulated, whereas in hyperthyroid rats the expression of Oatp1c1 is downregulated. Thus, the thyroid state-dependent regulation of Oatp1c1 counteracts the effects of alternations in circulating T4 levels on brain T4 uptake. Recently, Bronger et al. reported that they detected OATP1A2(OATP-A) expression in the apical membrane of human endothelial cells of the BBB by immunofluorescence microscopy (Bronger et al. 2005). On the other hand, no significant immunostaining was observed for OATP1C1(OATP-F) in human brain samples. So OATP1C1 (OATP-F) localization in human BBB is still controversial. OATP1A2 was reported that it transports T3 ($Km=6.5$ μM) and T4 ($Km=8.0$ μM) (Fujiwara et al. 2001) and is expressed exclusively in the brain (Abe et al. 1999). OATP1A2 might be also a component of the hormone transporting system in human BBB.

Recently, the generation and analysis of Oatp1c1-deficient mice were reported (Mayerl et al. 2012). Oatp1c1 knockout (KO) mice were grown without obvious growth retardation and neurological abnormalities. The concentration of serum T3 and T4 in Oatp1c1 KO animals was not significantly different from control group,

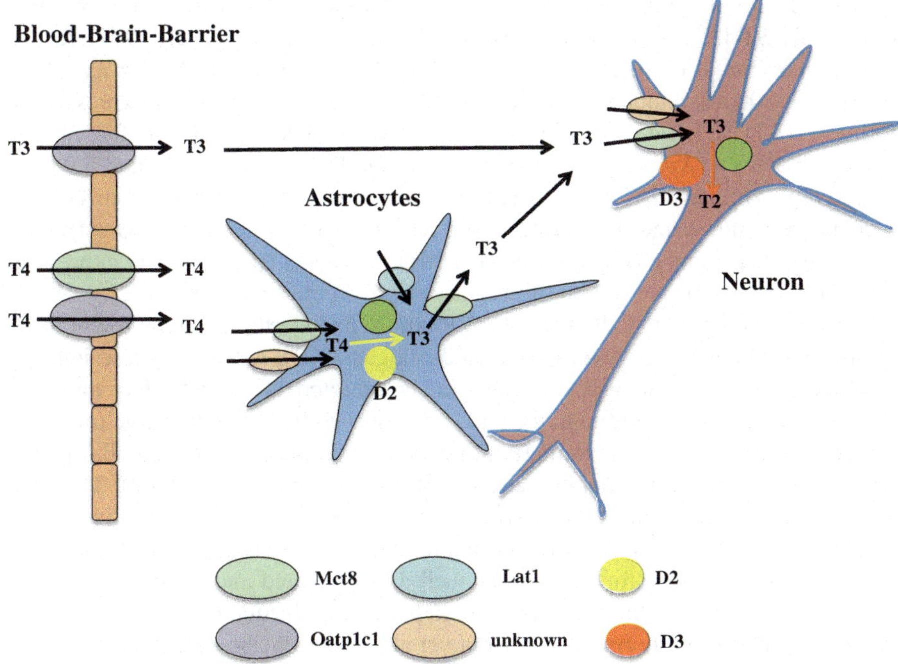

Fig. 11.1 The putative scheme of thyroid hormone transport in mouse brain. T4 is transported through blood–brain barrier (BBB) by Oapt1c1 or MCT8 into central nervous system (CNS). Transportation of T3 at BBB is mainly dependent on MCT8. T4 is uptake by still unknown transporting system in astrocytes. T4 is converted to active T3 by type2 deiodinase (D2) that is localized in astrocytes. T3 is uptake into neurons by MCT8 or unknown transporter(s). Inactivation of T3 is mediated by type3 deiodinase (D3) in neurons. System L amino acid transporter (Lat1) may be involved in T3 uptake into astrocytes

but the content of T3 and T4 in central nervous systems was decreased in Oatp1c1-deficient mice. Interestingly, increased type2 deiodinase and decreased type3 deiodinase activities in the brain of Oapt1c1 KO mice (Mayerl et al. 2012) were suggesting the relatively mild hypothyroid state in central nervous system of Oapt1c1-deficient animals and indicating the defective T4 transport across BBB due to the deficiency of functional Oatp1c1 in the KO mice brain (Fig. 11.1).

11.4   Several OATPs Are Involved in Thyroid Hormone Transport in Humans

Compared with rat Oatps, the expression patterns of human organic anion transporters are more tissue specific. Human OATP1A2(OATP-A), which transports T3 ($Km=6.5$ μM) and T4 ($Km=8.0$ μM) (Fujiwara et al. 2001), is expressed exclusively in the brain (Abe et al. 1999). Human liver-specific transporters OATP1B1(LST-1/OATP-C/OATP2) (Abe et al. 1999; Hsiang et al. 1999; Konig et al. 2000a; Tamai et al. 2000) and OATP1B3(LST-2/OATP8) (Abe et al. 2001; Konig et al. 2000b), which transport thyroid hormone, are exclusively expressed in the liver. OATP1B1 transports thyroid hormone with a Km of 3.0 μM for T4 and 2.7 μM for T3. The Km values for T4 and T3 of OATP1B3 are 6.5 μM and 5.9 μM, respectively. Human OATP4A1(OATP-E) transports thyroid hormone in a Na^+-independent manner, and the Km value for T3 is 0.9 μM, which is the lowest value among the Oatp family (Fujiwara et al. 2001). OATP4A1(OATP-E) is expressed abundantly in various peripheral tissues, as well as in the small intestine.

It is well known that study of the OATP family had started from the research of the bile acid uptake in the liver (Meier and Stieger 2002). After biosynthesis from the cholesterol in the hepatocyte, bile acid is excreted in the bile duct and this secreted bile acid in the alimentary canal is reabsorbed from the digestive tract and reentered to the blood. This circulating bile acid is re-uptaked by liver cells. Those processes are called enterohepatic circulation (Meier and Stieger 2002). The first member of organic anion transporter family identified in the rat liver was Oatp1a1(oatp1), the Na^+-independent organic anion transporting polypeptide, transports bile acid, bromosulfophthalein, as well as conjugated and unconjugated steroid hormones from the blood stream into hepatocytes (Jacquemin et al. 1994) and also transports thyroid hormone (Friesema et al. 1999). Physiological studies have suggested the presence of other members of this family in the liver. Subsequently, Oatp1a4(oatp2) and Oatp1a5(oatp3) in rat, OATP1B1(LST-1/OATP-C/OATP2) and OATP1B3(LST-2/OATP8) in human are identified as bile acid transporters.

The enterohepatic circulation of thyroid hormone has also already been well established in rat (Albert and Keating 1952; DiStefano et al. 1992). Because OATP4A1, OATP1B1, and OATP1B3 transport both thyroid hormone and taurocholate, it is suggested that, in human, OATP4A1, OATP1B1, and OATP1B3 might be involved in the transport of bile acids and thyroid hormone in the enterohepatic circulation. These data suggest the possibility that molecules, which transport bile acid, also transport thyroid hormone in vivo.

Rat gonad-specific organic anion transporter Oatp6b1(GST-1) and Oatp6c1(GST-2) have been isolated and are abundantly expressed in testis and moderately in ovary, adrenal gland, and epididymis (Suzuki et al. 2003). Both Oatp6b1 and Oatp6c1 are members of the oatp family and are reported to transport T4 with Km of 6.4 μM and 5.8 μM, respectively. The human counterpart, human GST, was also isolated and is exclusively expressed in the testis (Suzuki et al. 2003). Because hypothyroidism in neonatal and early infantile periods impairs the development and maturation of gonads and causes infertility in adults, those gonad-specific transporters might be involved in the regulation of the development and function of gonads.

In the kidney, kidney-specific organic anion transporter OATP4C1(OATP-R) and its rat counterpart Oatp4c1(oatp-R) have been isolated and proven to transport both T3 and T4 (Toyohara et al. 2009). The Km values for T3 of human OATP4C1 and rat Oatp4c1 are 5.9 μM and 1.9 μM, respectively. Immunohistochemistry revealed that rat Oatp4c1 was expressed in the proximal tubules and so far no other

OATP family member that transports thyroid hormone was found to be expressed in human kidney. In addition, the proximal tubule cells express type1 5′-deiodinase, which is a major enzyme catalyzing the conversion of T3 from T4 (Lee et al. 1993). Colocalization of Oatp4c1 and type1 5′-deiodinase might be functionally concerned in the regulation of thyroid hormone metabolism in the proximal tubules.

11.5 Monocarboxylate Transporter Family

The MCT(SLC16) family is named because the first four members have been characterized as transporters of monocarboxylates such as lactate, pyruvate, and ketone bodies (Halestrap 2013). To date, 14 members of the MCT family have been identified in various tissues from different species. However, the types of ligands have been identified for only six members (i.e., monocarboxylates for MCT1-4 and aromatic amino acid derivatives for MCT8 and MCT10). The functions of the other MCT families remain to be determined. The MCTs are proteins of 426–613 amino acids with 12 predicted transmembrane domains. Among MCTs, MCT8 (SLC16As) is expressed in many tissues, including human liver, kidney, heart, brain, placenta, lung, and skeletal muscle (Friesema et al. 2003). MCT8 is located on X chromosome (Xq13.2), comprised of six exons and containing two alternative translation start sites that code for predicted 613 and 539 amino acids proteins, respectively (Halestrap 2013; Friesema et al. 2006a; Bernal et al. 2015). Recently, Friesema et al. identified rat MCT8 as a specific thyroid hormone transporter (Friesema et al. 2003). MCT8 expression in *Xenopus* oocytes showed the uptake of T4, T3, rT3, and 3,3′-T2, and the *Km* values of 4.7 M for T4, 4.0 M for T3, and 2.2 M for rT3, which was at an order lower than for the isolated other thyroid hormone transporters. This transport of T3 by MCT8 is Na$^+$-independent, although T4 transport is decreased in the absence of Na$^+$ (Friesema et al. 2003). In mouse central nervous system, the highest mRNA expression levels of MCT8 were detected in neuronal cells of the cerebrocortex, olfactory bulb, hippocampus, and amygdala; moderate levels in the striatum and cerebellum; and low levels in some neuroendocrine nuclei (Heuer et al. 2005). These neural cell populations are thyroid hormone-sensitive and the thyroid hormone transport system in those neural tissues might be mediated by MCT8 (Fig. 11.1).

The pathophysiological relevance of MCT8 as a thyroid hormone transporter has been reported in patients with a novel syndrome of severe X-linked mental retardation and psychomotor disability and strongly elevated T3 levels and low T4 (Friesema et al. 2004, 2006b; Dumitrescu et al. 2004; Jansen et al. 2005). This X-linked mental retardation and psychomotor disability has been know as Allan-Herndon-Dudley syndrome (AHDS) (Schwartz et al. 2005). This severe phenotype is explained partly by the recent reports that MCT8 is expressed specifically in thyroid hormone sensitive neuronal populations (Heuer et al. 2005; Alkemade et al. 2005, 2006). MCT8 might be important for the neuronal uptake of T3 produced from T4 in neighboring astrocytes that express type2 iodothyronine deiodinase, which catalyzes the conversion of T3 from T4 (Guadano-Ferraz et al. 1997) (see Fig. 11.1).

Two research groups have reported MCT8-deficient mice (Dumitrescu et al. 2006; Trajkovic et al. 2007). MCT8 knockout mice show disturbed profiles of serum thyroid hormone levels, high concentration of serum T3 and low serum T4 level, however there was no obvious phenotype of developmental retardation in nervous system (Dumitrescu et al. 2006; Trajkovic et al. 2007). The fact that the neurological manifestations present in MCT8-deficient human have not been replicated in the mouse model might be due to species-specific regulation of intracellular thyroid hormone availability or demands of thyroid hormone for normal function in central nervous system.

Uptake of T3 into the brain was almost absent in MCT8-null mice, suggests that MCT8 is a key molecule in T3 transverse into the brain (Dumitrescu et al. 2006; Trajkovic et al. 2007) (Fig. 11.1). Whereas, entry of T4 into the brain was not changed in the absence of MCT8 (Trajkovic et al. 2007), indicates the presence of other thyroid hormone transporters such as Oatp1c1 at the BBB that which carry T4 into the tissues (Tohyama et al. 2004; Roberts et al. 2008) (Fig. 11.1). Trajkovic et al. reported that there was no obvious morphological abnormality of Purkinje cells and the external granule cell layer in the cerebellum of the 12-day-old MCT8-deficient mice compared to control animals (Trajkovic et al. 2007). Purkinje cell cultures from both neonatal MCT8-null and wild-type mice demonstrated the similar normal neuronal sensitivity (increased dendritic parameters) to T3 treatment in vitro (Trajkovic et al. 2007), implies the neuronal MCT8 deficiency might be compensated for by other thyroid hormone transporters in mice, and MCT8-null mice are protected from the neurological deficit observed in humans.

Recently, mice deficient of both MCT8 and Oapt1c1 were developed and characterized (Mayerl et al. 2014). MCT8 and Oatp1c1 double knockout mice (Mct8/Oatp1c1 DKO) displayed more severely decreased uptakes and contents of both T3 and T4 compared with either Mct8 or Oatp1c1 single KO animals. Probably due to markedly decreasing T3 and T4 content in brain, diminished expressions of thyroid hormone target genes (aldehyde dehydrogenase 1a1, neurogranin and hairless) as well as enhanced activity of type2 iodinase (responsible enzyme for conversion of prohormone T4 to more bioactive hormone T3) were observed. As expected from the more severely impaired utilization of thyroid hormones in CNS compared with Mct8 KO or Oatp1c1 KO animals, Mct8/Oatp1c1 DKO mice exhibited delayed cerebellar development, reduced myelination, and abnormal locomotor activities (Mayerl et al. 2014).

11.6 Amino Acid Transporter Is Member of Thyroid Hormone Transporter Family

Amino acids, especially essential amino acids, are required for protein synthesis and as energy sources in all living cells. Because most amino acids are hydrophilic, they require special membrane transport systems to penetrate the cell membrane. Many amino acid transporters, each of which have 12 transmembrane regions, have been characterized (Malandro and Kilberg 1996). In addition, the cell-surface glycoprotein 4F2 heavy chain (4F2hc; CD98 in mouse) has also been identified as being part of an amino acid transporter. 4F2hc has a single transmembrane domain and a 120-kDa disulfide-linked protein formed by a heteromeric glycosylated 40 kDa light chain together with an 80 kDa heavy chain (Haynes et al. 1981; Mastroberardino et al. 1998). The expression of 4F2hc in *Xenopus* oocytes induces low levels of cysteine, dibasic and neutral amino acid transport (Wells et al. 1992). Friesema et al. (2001) reported that co-injection of cRNAs for 4F2hc and human LAT1 light chain not only transported amino acids (Phe, Tyr, Leu, and Trp), which is a characteristic of the system L amino acid transporter (a Na^+-independent exchange of large, neutral amino acids), but also transported thyroid hormone in a Na^+-independent manner. The *Km* values for T4, T3, reverse T3, and T2 were 7.9 μM, 0.8 μM, 12.5 μM, and 7.9 μM, respectively. Bloundeau et al. (1993) showed that aromatic and neutral amino acids are transported into cultured astrocytes via the Na^+-independent system L, which is related to the thyroid hormone transport system. These data suggest the contribution of the amino acid transporter LAT-1/4F2hc complex to thyroid hormone transport in astrocytes (Fig. 11.1).

11.7 Differences of *Km* Values and Na^+ dependency

Several questions arise from recent work on the identification of thyroid hormone transporters, the most common being why are there different *Km* values between cDNA expression systems and in vitro culture systems? The in vitro studies characterized two saturable sites for thyroid hormone binding; one is characterized by high affinity and low capacity, and the other by low affinity and high capacity (Krenning et al. 1981; Hennemann et al. 2001). The high-affinity uptake of T3 is mostly in the nM range and is often energy and extracellular Na^+ dependent. However, in the low-affinity uptake process the *Km* for T4 and T3 of the fraction is in the M range, which is similar to that of oatps and amino acid transporters expressed in *Xenopus* oocytes (Friesema et al. 2001), hepatocytes (Blondeau et al. 1988), and glial cells (Francon et al. 1989). Little is known about the molecular entity responsible for the high affinity-system, except OATP1C1(OATP-F) and Oatp1c1(oatp14), although differences in the types of assays or experimental procedures used are possible explanations for the discrepant results.

Another controversy arises from the fact that the oatp family- and amino acid transporter-mediated uptake of thyroid hormone is not dependent on extracellular Na^+ in vitro, whereas the transporting mechanisms of thyroid hormones in the tissues are heterogeneous in terms of Na^+ dependency: Na^+-dependent (Blondeau et al. 1988), Na^+-independent (Topliss et al. 1989; Centanni and Robbins 1987), and mixed (Francon et al. 1989; Galton et al. 1986). Rat ntcp was reported to transport thyroid hormone (Friesema et al. 1999). Ntcp consists of ~350 amino acids with seven putative transmembrane domains (Hagenbuch et al. 1991). Ntcp is expressed only in the basolateral cell membrane of hepatocytes and is the major transporting mechanism for conjugated bile acids, especially taurocholate, into the liver (other bile acids are mainly transported by oatps) (Hagenbuch and Dawson 2004). The Na$^+$-independent fraction of thyroid hormone uptake into the tissues could be partly attributed to the oatp family and 4F2c-L-type amino acid transporter complex. However, the molecular entities responsible for the Na$^+$-dependent system in various tissues are still unclear, because the Na$^+$-dependent transporters, ntcp and ASBT, are not expressed widely (Hagenbuch and Dawson 2004). There might be several other candidate molecules that are responsible for transporting thyroid hormone.

11.8 Thyroid Hormone Transport Related Diseases

Certain individuals exhibit a syndrome of resistance to thyroid hormone in many tissues. Such patients have reduced clinical and biochemical manifestations thyroid hormone action relative to the circulating hormone level (Resistance to thyroid hormone, RTH, or Refetoff's syndrome) (Refetoff et al. 1993; Yen 2003). In practice, most patients are identified by the persistent elevation of serum levels of T4 and T3 with inappropriately non-suppressed TSH, in the absence of intercurrent acute illness, drugs, or alteration of thyroid hormone binding to serum proteins. One well-known molecular basis is an abnormality of the nuclear thyroid hormone receptor (Yen 2003). However, other functional defects in such patients have also been postulated (Refetoff et al. 1993). One possible mechanism is reduced hormone availability to tissues because of impaired thyroid hormone entry into target cells. Thus, the finding of a thyroid hormone transporter might provide a clue for identifying the genetic basis for the lack of thyroid hormone responsiveness.

References

Abe T, Kakyo M, Sakagami H, Tokui T, Nishio T, Tanemoto M et al (1998) Molecular characterization and tissue distribution of a new organic anion transporter subtype (oatp3) that transports thyroid hormones and taurocholate and comparison with oatp2. J Biol Chem 273(35):22395–22401

Abe T, Kakyo M, Tokui T, Nakagomi R, Nishio T, Nakai D et al (1999) Identification of a novel gene family encoding human liver-specific organic anion transporter LST-1. J Biol Chem 274(24):17159–17163

Abe T, Unno M, Onogawa T, Tokui T, Kondo TN, Nakagomi R et al (2001) LST-2, a human liver-specific organic anion transporter, determines methotrexate sensitivity in gastrointestinal cancers. Gastroenterology 120(7):1689–1699

Abe T, Suzuki T, Unno M, Tokui T, Ito S (2002) Thyroid hormone transporters: recent advances. Trends Endocrinol Metab 13(5):215–220

Albert A, Keating FR Jr (1952) The role of the gastrointestinal tract, including the liver, in the metabolism of radiothyroxine. Endocrinology 51(5):427–443

Alkemade A, Friesema EC, Unmehopa UA, Fabriek BO, Kuiper GG, Leonard JL et al (2005) Neuroanatomical pathways for thyroid hormone feedback in the human hypothalamus. J Clin Endocrinol Metab 90(7):4322–4334

Alkemade A, Friesema EC, Kuiper GG, Wiersinga WM, Swaab DF, Visser TJ et al (2006) Novel neuroanatomical pathways for thyroid hormone action in the human anterior pituitary. Eur J Endocrinol 154(3):491–500

Angeletti RH, Novikoff PM, Juvvadi SR, Fritschy JM, Meier PJ, Wolkoff AW (1997) The choroid plexus epithelium is the site of the organic anion transport protein in the brain. Proc Natl Acad Sci U S A 94(1):283–286

Balazs R, Brooksbank BW, Davison AN, Eayrs JT, Wilson DA (1969) The effect of neonatal thyroidectomy on myelination in the rat brain. Brain Res 15(1):219–232

Balazs R, Kovacs S, Cocks WA, Johnson AL, Eayrs JT (1971) Effect of thyroid hormone on the biochemical maturation of rat brain: postnatal cell formation. Brain Res 25(3):555–570

Bernal J, Guadano-Ferraz A, Morte B (2015) Thyroid hormone transporters—functions and clinical implications. Nat Rev Endocrinol 11(7):406–417

Beslin A, Chantoux F, Blondeau JP, Francon J (1995) Relationship between the thyroid hormone transport system and the Na(+)-H+ exchanger in cultured rat brain astrocytes. Endocrinology 136(12):5385–5390

Blondeau JP, Osty J, Francon J (1988) Characterization of the thyroid hormone transport system of isolated hepatocytes. J Biol Chem 263(6):2685–2692

Blondeau JP, Beslin A, Chantoux F, Francon J (1993) Triiodothyronine is a high-affinity inhibitor of amino acid transport system L1 in cultured astrocytes. J Neurochem 60(4):1407–1413

Bronger H, Konig J, Kopplow K, Steiner HH, Ahmadi R, Herold-Mende C et al (2005) ABCC drug efflux pumps and organic anion uptake transporters in human gliomas and the blood-tumor barrier. Cancer Res 65(24):11419–11428

Cavallaro T, Martone RL, Dwork AJ, Schon EA, Herbert J (1990) The retinal pigment epithelium is the unique site of transthyretin synthesis in the rat eye. Invest Ophthalmol Vis Sci 31(3):497–501

Centanni M, Robbins J (1987) Role of sodium in thyroid hormone uptake by rat skeletal muscle. J Clin Invest 80(4):1068–1072

Chanoine JP, Alex S, Fang SL, Stone S, Leonard JL, Korhle J et al (1992) Role of transthyretin in the transport of thyroxine from the blood to the choroid plexus, the cerebrospinal fluid, and the brain. Endocrinology 130(2):933–938

Chantoux F, Blondeau JP, Francon J (1995) Characterization of the thyroid hormone transport system of cerebrocortical rat neurons in primary culture. J Neurochem 65(6):2549–2554

Chatonnet F, Flamant F, Morte B (2015) A temporary compendium of thyroid hormone target genes in brain. Biochim Biophys Acta 1849(2):122–129

DiStefano JJ III, Nguyen TT, Yen YM (1992) Sites and patterns of absorption of 3,5,3′-triiodothyronine and thyroxine along rat small and large intestines. Endocrinology 131(1):275–280

Dumitrescu AM, Liao XH, Best TB, Brockmann K, Refetoff S (2004) A novel syndrome combining thyroid and neurological abnormalities is associated with mutations in a monocarboxylate transporter gene. Am J Hum Genet 74(1):168–175

Dumitrescu AM, Liao XH, Weiss RE, Millen K, Refetoff S (2006) Tissue-specific thyroid hormone deprivation and excess in monocarboxylate transporter (mct) 8-deficient mice. Endocrinology 147(9):4036–4043

Eayrs JT (1960) Influence of the thyroid on the central nervous system. Br Med Bull 16:122–127

Francon J, Chantoux F, Blondeau JP (1989) Carrier-mediated transport of thyroid hormones into rat glial cells in primary culture. J Neurochem 53(5):1456–1463

Friesema EC, Docter R, Moerings EP, Stieger B, Hagenbuch B, Meier PJ et al (1999) Identification of thyroid hormone transporters. Biochem Biophys Res Commun 254(2):497–501

Friesema EC, Docter R, Moerings EP, Verrey F, Krenning EP, Hennemann G et al (2001) Thyroid hormone transport by the heterodimeric human system L amino acid transporter. Endocrinology 142(10):4339–4348

Friesema EC, Ganguly S, Abdalla A, Manning Fox JE, Halestrap AP, Visser TJ (2003) Identification of monocarboxylate transporter 8 as a specific thyroid hormone transporter. J Biol Chem 278(41):40128–40135

Friesema EC, Grueters A, Biebermann H, Krude H, von Moers A, Reeser M et al (2004) Association between mutations in a thyroid hormone transporter and severe X-linked psychomotor retardation. Lancet 364(9443):1435–1437

Friesema EC, Jansen J, Visser TJ (2005) Thyroid hormone transporters. Biochem Soc Trans 33(Pt 1):228–232

Friesema EC, Kuiper GG, Jansen J, Visser TJ, Kester MH (2006a) Thyroid hormone transport by the human monocarboxylate transporter 8 and its rate-limiting role in intracellular metabolism. Mol Endocrinol 20(11):2761–2772

Friesema EC, Jansen J, Heuer H, Trajkovic M, Bauer K, Visser TJ (2006b) Mechanisms of disease: psychomotor retardation and high T3 levels caused by mutations in monocarboxylate transporter 8. Nat Clin Pract Endocrinol Metab 2(9):512–523

Fujiwara K, Adachi H, Nishio T, Unno M, Tokui T, Okabe M et al (2001) Identification of thyroid hormone transporters in humans: different molecules are involved in a tissue-specific manner. Endocrinology 142(5):2005–2012

Galton VA, St Germain DL, Whittemore S (1986) Cellular uptake of 3,5,3′-triiodothyronine and thyroxine by red blood and thymus cells. Endocrinology 118(5):1918–1923

Gao B, Stieger B, Noe B, Fritschy JM, Meier PJ (1999) Localization of the organic anion transporting polypeptide 2 (Oatp2) in capillary endothelium and choroid plexus epithelium of rat brain. J Histochem Cytochem 47(10):1255–1264

Gingrich SA, Smith PJ, Shapiro LE, Surks MI (1985) 5,5′-Diphenylhydantoin (phenytoin) attenuates the action of 3,5,3′-triiodo-L-thyronine in cultured GC cells. Endocrinology 116(6):2306–2313

Guadano-Ferraz A, Obregon MJ, St Germain DL, Bernal J (1997) The type 2 iodothyronine deiodinase is expressed primarily in glial cells in the neonatal rat brain. Proc Natl Acad Sci U S A 94(19):10391–10396

Hagenbuch B, Dawson P (2004) The sodium bile salt cotransport family SLC10. Pflugers Arch 447(5):566–570

Hagenbuch B, Stieger B, Foguet M, Lubbert H, Meier PJ (1991) Functional expression cloning and characterization of the hepatocyte Na+/bile acid cotransport system. Proc Natl Acad Sci U S A 88(23):10629–10633

Halestrap AP (2013) The SLC16 gene family—structure, role and regulation in health and disease. Mol Aspects Med 34(2–3):337–349

Haynes BF, Hemler ME, Mann DL, Eisenbarth GS, Shelhamer J, Mostowski HS et al (1981) Characterization of a monoclonal antibody (4F2) that binds to human monocytes and to a subset of activated lymphocytes. J Immunol 126(4):1409–1414

Hennemann G, Docter R, Friesema EC, de Jong M, Krenning EP, Visser TJ (2001) Plasma membrane transport of thyroid hormones and its role in thyroid hormone metabolism and bioavailability. Endocr Rev 22(4):451–476

Heuer H, Maier MK, Iden S, Mittag J, Friesema EC, Visser TJ et al (2005) The monocarboxylate transporter 8 linked to human psychomotor retardation is highly expressed in thyroid hormone-sensitive neuron populations. Endocrinology 146(4):1701–1706

Hsiang B, Zhu Y, Wang Z, Wu Y, Sasseville V, Yang WP et al (1999) A novel human hepatic organic anion transporting polypeptide (OATP2). Identification of a liver-specific human organic anion transporting polypeptide and identification of rat and human hydroxymethylglutaryl-CoA reductase inhibitor transporters. J Biol Chem 274(52):37161–37168

Ito A, Yamaguchi K, Onogawa T, Unno M, Suzuki T, Nishio T et al (2002) Distribution of organic anion-transporting polypeptide 2 (oatp2) and oatp3 in the rat retina. Invest Ophthalmol Vis Sci 43(3):858–863

Jacquemin E, Hagenbuch B, Stieger B, Wolkoff AW, Meier PJ (1994) Expression cloning of a rat liver Na(+)-independent organic anion transporter. Proc Natl Acad Sci U S A 91(1):133–137

Jansen J, Friesema EC, Milici C, Visser TJ (2005) Thyroid hormone transporters in health and disease. Thyroid 15(8):757–768

Konig J, Cui Y, Nies AT, Keppler D (2000a) A novel human organic anion transporting polypeptide localized to the basolateral hepatocyte membrane. Am J Physiol Gastrointest Liver Physiol 278(1):G156–G164

Konig J, Cui Y, Nies AT, Keppler D (2000b) Localization and genomic organization of a new hepatocellular organic anion transporting polypeptide. J Biol Chem 275(30):23161–23168

Krenning E, Docter R, Bernard B, Visser T, Hennemann G (1981) Characteristics of active transport of thyroid hormone into rat hepatocytes. Biochim Biophys Acta 676(3):314–320

Kusuhara H, He Z, Nagata Y, Nozaki Y, Ito T, Masuda H et al (2003) Expression and functional involvement of organic anion transporting polypeptide subtype 3 (Slc21a7) in rat choroid plexus. Pharm Res 20(5):720–727

Lee WS, Berry MJ, Hediger MA, Larsen PR (1993) The type I iodothyronine 5′-deiodinase messenger ribonucleic acid is localized to the S3 segment of the rat kidney proximal tubule. Endocrinology 132(5):2136–2140

Malandro MS, Kilberg MS (1996) Molecular biology of mammalian amino acid transporters. Annu Rev Biochem 65:305–336

Mastroberardino L, Spindler B, Pfeiffer R, Skelly PJ, Loffing J, Shoemaker CB et al (1998) Amino-acid transport by heterodimers of 4F2hc/CD98 and members of a permease family. Nature 395(6699):288–291

Mayerl S, Visser TJ, Darras VM, Horn S, Heuer H (2012) Impact of Oatp1c1 deficiency on thyroid hormone metabolism and action in the mouse brain. Endocrinology 153(3):1528–1537

Mayerl S, Muller J, Bauer R, Richert S, Kassmann CM, Darras VM et al (2014) Transporters MCT8 and OATP1C1 maintain murine brain thyroid hormone homeostasis. J Clin Invest 124(5):1987–1999

Meier PJ, Stieger B (2002) Bile salt transporters. Annu Rev Physiol 64:635–661

Ohtsuki S, Takizawa T, Takanaga H, Terasaki N, Kitazawa T, Sasaki M et al (2003) In vitro study of the functional expression of organic anion transporting polypeptide 3 at rat choroid plexus epithelial cells and its involvement in the cerebrospinal fluid-to-blood transport of estrone-3-sulfate. Mol Pharmacol 63(3):532–537

Ohtsuki S, Takizawa T, Takanaga H, Hori S, Hosoya K, Terasaki T (2004) Localization of organic anion transporting polypeptide 3 (oatp3) in mouse brain parenchymal and capillary endothelial cells. J Neurochem 90(3):743–749

Oppenheimer JH, Schwartz HL (1997) Molecular basis of thyroid hormone-dependent brain development. Endocr Rev 18(4):462–475

Osty J, Jego L, Francon J, Blondeau JP (1988) Characterization of triiodothyronine transport and accumulation in rat erythrocytes. Endocrinology 123(5):2303–2311

Palha JA, Fernandes R, de Escobar GM, Episkopou V, Gottesman M, Saraiva MJ (2000) Transthyretin regulates thyroid hormone levels in the choroid plexus, but not in the brain parenchyma: study in a transthyretin-null mouse model. Endocrinology 141(9):3267–3272

Pizzagalli F, Hagenbuch B, Stieger B, Klenk U, Folkers G, Meier PJ (2002) Identification of a novel human organic anion transporting polypeptide as a high affinity thyroxine transporter. Mol Endocrinol 16(10):2283–2296

Porterfield SP, Hendrich CE (1993) The role of thyroid hormones in prenatal and neonatal neurological development—current perspectives. Endocr Rev 14(1):94–106

Refetoff S, Weiss RE, Usala SJ (1993) The syndromes of resistance to thyroid hormone. Endocr Rev 14(3):348–399

Roberts LM, Woodford K, Zhou M, Black DS, Haggerty JE, Tate EH et al (2008) Expression of the thyroid hormone transporters monocarboxylate transporter-8 (SLC16A2) and organic ion transporter-14 (SLCO1C1) at the blood-brain barrier. Endocrinology 149(12):6251–6261

Schwartz CE, May MM, Carpenter NJ, Rogers RC, Martin J, Bialer MG et al (2005) Allan-Herndon-Dudley syndrome and the MCT8 thyroid hormone transporter. Am J Hum Genet. 77(1):41–53

Southwell BR, Duan W, Alcorn D, Brack C, Richardson SJ, Kohrle J et al (1993) Thyroxine transport to the brain: role of protein synthesis by the choroid plexus. Endocrinology 133(5):2116–2126

Sugiyama D, Kusuhara H, Taniguchi H, Ishikawa S, Nozaki Y, Aburatani H et al (2003) Functional characterization of rat brain-specific organic anion transporter (Oatp14) at the blood-brain barrier: high affinity transporter for thyroxine. J Biol Chem 278(44):43489–43495

Suzuki T, Onogawa T, Asano N, Mizutamari H, Mikkaichi T, Tanemoto M et al (2003) Identification and characterization of novel rat and human gonad-specific organic anion transporters. Mol Endocrinol 17(7):1203–1215

Tamai I, Nezu J, Uchino H, Sai Y, Oku A, Shimane M et al (2000) Molecular identification and characterization of novel members of the human organic anion transporter (OATP) family. Biochem Biophys Res Commun 273(1):251–260

Thompson CC, Potter GB (2000) Thyroid hormone action in neural development. Cereb Cortex 10(10):939–945

Tohyama K, Kusuhara H, Sugiyama Y (2004) Involvement of multispecific organic anion transporter, Oatp14 (Slc21a14), in the transport of thyroxine across the blood-brain barrier. Endocrinology 145(9):4384–4391

Topliss DJ, Kolliniatis E, Barlow JW, Lim CF, Stockigt JR (1989) Uptake of 3,5,3′-triiodothyronine by cultured rat hepatoma cells is inhibitable by nonbile acid cholephils, diphenylhydantoin, and nonsteroidal antiinflammatory drugs. Endocrinology 124(2):980–986

Toyohara T, Suzuki T, Morimoto R, Akiyama Y, Souma T, Shiwaku HO et al (2009) SLCO4C1 transporter eliminates uremic toxins and attenuates hypertension and renal inflammation. J Am Soc Nephrol 20(12):2546–2555

Trajkovic M, Visser TJ, Mittag J, Horn S, Lukas J, Darras VM et al (2007) Abnormal thyroid hormone metabolism in mice lacking the monocarboxylate transporter 8. J Clin Invest 117(3):627–635

Wells RG, Lee WS, Kanai Y, Leiden JM, Hediger MA (1992) The 4F2 antigen heavy chain induces uptake of neutral and dibasic amino acids in Xenopus oocytes. J Biol Chem 267(22):15285–15288

Yen PM (2001) Physiological and molecular basis of thyroid hormone action. Physiol Rev 81(3):1097–1142

Yen PM (2003) Molecular basis of resistance to thyroid hormone. Trends Endocrinol Metab 14(7):327–333

Chapter 12
Syndromes of Resistance to Thyroid Hormone and Brain Development

Irene Campi and Paolo Beck-Peccoz

Abstract Thyroid hormones (TH) are crucial for neuronal development, as untreated congenital hypothyroidism or genetic defects of TH transport cause severe neurological and cognitive impairment (Bernal, Nat Clin Pract Endocrinol Metab 3:249–259, 2007). TH modulate the division, the cell cycle, the migration, the maturation, and the differentiation of neuronal progenitors.

A number of syndromes are associated with reduced responsiveness to thyroid hormones, expanding the original definition of thyroid hormone resistance, firstly described by Refetoff and collaborators in 1967, which is characterized by elevated circulating levels of T4 and T3 with measurable serum TSH concentrations, as a consequence of mutations of thyroid hormone receptor beta (TRβ) (Refetoff et al., J Clin Endocr 27:279–294, 1967). Nowadays, other forms of insensitivity to TH have been identified: defects in cell surface transporters such as the monocarboxylate transporter 8 (MCT8), genetic disorder of thyroid hormone metabolism due to alterations of selenoproteins synthesis, which comprise the deiodinase enzymes, and finally, mutations in the thyroid hormone receptor alpha (TRα) (Refetoff et al., Endocr Rev 14:348–399, 1993; Refetoff and Dumitrescu, Best Pract Res Clin Endocrinol Metab 21:277–305, 2007; Agrawal et al., Postgrad Med J 84:473–477, 2008; Gurnell et al., Endocrinology, adult and pediatric, Sauderns Elsevier, Philadelphia, pp 1745–1759, 2010; Refetoff and Dumitrescu, http://www.thyroidmanager.org, 2010; Cheng et al., Endocr Rev 31:139–170, 2010; Visser et al., Mol Endocrinol 25:1–14, 2011; Hammes et al., Cell 122:751–762, 2005; Schoenmakers et al., J Clin Invest 120:4220–4235, 2010; Bochukova et al., N Engl J Med 366:243–249, 2012; van Mullem et al., N Engl J Med 366:1451–1453, 2012; Schoenmakers et al., Biochim Biophys Acta 1830:4004–4008, 2013; Moran et al., J Clin Endocrinol Metab 98:4254–4261, 2013).

I. Campi • P. Beck-Peccoz, M.D. (✉)
Department of Clinical Sciences and Community Health, University of Milan, Fondazione IRCCS Ca' Granda Ospedale Maggiore, Pad. Granelli, Via F. Sforza 35, Milan 20122, Italy
e-mail: paolo.beckpeccoz@unimi.it

© Springer Science+Business Media New York 2016
N. Koibuchi, P.M. Yen (eds.), *Thyroid Hormone Disruption and Neurodevelopment*,
Contemporary Clinical Neuroscience, DOI 10.1007/978-1-4939-3737-0_12

In this chapter we will describe the neurological phenotype associated with these syndromes, with the exception of the inherited defects in thyroid hormone transporters such as the monocarboxylate transporter 8 (MCT-8), described elsewhere in this book.

Keywords TRβ • Thyroid response elements (TRE) • RTH β • RTH α • GRTH • PRTH • TSH-oma • Glycoprotein hormone α-subunit (α-GSU) • SHBG • Deiodinases • Selenoproteins • SECISBP2

12.1 Thyroid Hormone Actions in Brain and Other Target Tissues

The biological actions of TH in target tissues are regulated by a number of variables such as the deiodinases system, which controls the metabolism of TH, the cell surface transporters, and thyroid hormone receptors, which in concert with different cofactors, modulate the TH nuclear actions (Cheng et al. 2010). Dysfunction of any of these regulatory step may cause a reduced responsiveness to TH at peripheral level. The known human syndromes of resistance to TH and their biochemical hallmarks are summarized in Table 12.1. Defects in TH serum transport proteins, such as thyroxine binding globulin (TBG) and albumin may result in spuriously high levels of FT4 and/or FT3 and should be considered in the differential diagnosis of these syndromes (Table 12.1).

12.1.1 Deiodination System

The intracellular concentrations of T3 are regulated by three deiodinases, selenocysteine-containing enzymes, which activate or inactivate the iodothyronines through the removal of iodide from thyroxine and its metabolites.

The incorporation of the selenium (Se) in the selenoproteins is mediated by a multiprotein complex; genetic mutations in proteins belonging to this system, such as the selenocysteine insertion sequence-binding protein 2 (SECIBP2), cause a deficient production of selenoprotein associated with an abnormal thyroid hormone metabolism (Refetoff and Dumitrescu 2007; Schoenmakers et al. 2010).

Type 1 deiodinase (DIO1) deiodinate T4 (3,5,3′,5′-tetraiodothyronine) into T3 (3,5,3′-triiodothyronine) and of T3 into T2 (3,3′-diiodothyronine); DIO1 is highly expressed in liver and kidney and contribute to the production of plasma T3 by deiodination of T4. The type 2 deiodinase (DIO2) converts T4 to T3 intracellularly; it is highly expressed in the thyroid, being responsible for the increased intrathyroidal production of T3 in Graves' disease and toxic thyroid nodules.

Finally, type 3 deiodinase (DIO3) degrades T4 to rT3 and T3 to 3,3′-diiodothyronine (T2), thus downregulating the local T3 production and protecting tissues from TH excess.

Table 12.1 Genetic disorders characterized by increased serum thyroid hormones levels and detectable TSH concentrations

	Gene	Free T4	Free T3	TSH	Total reverse T3	SHBG[a]
Familial dysalbuminemic hyperthyroxinemia (FDH)	*ALB*	N[b]	N[b]	N	↑	N
Resistance to thyroid hormone (RTH β)	*THRB*	↑	↑	N or slightly ↑	↑	N
Defect of THRA gene (RTH α)	*THRA*	N/low end of normal range	N or slightly ↑	N	↓	↑
Defect of thyroid hormones transport (Allan–Herndon–Dudley syndrome)	*MCT8*	N or borderline ↓	↑	N or slightly ↑	↓	↑
Defect of thyroid hormones metabolism (SBP2 deficiency)	*SBP2*	↑	N or borderline ↓	N or slightly ↑	↑	N

[a]*SHBG* sex hormone-binding globulin
[b]As measured by equilibrium dialysis or direct "two-step" measurement methods. Interferences leading to spuriously high levels of FT4 and/or FT3 may be present by using other methods

In the CNS, DIO2 is mainly expressed in glial-derived cells, such as the astrocytes and the tanycytes, while DIO3 is primarily found in neurons. In order to maintain adequate levels of T3 in neural tissues, the deiodinases function is finely regulated. In particular, recent data suggest that astrocytes produce active T3, which enters in neurons via MCT-8. In these cells, DIO3 controls the local bioavailability of T3 by the production of rT3 and T2 from T4 and T3, respectively (Schroeder and Privalsky 2014).

12.1.2 Cell Surface Transporters

Since TH essentially exert their effects inside the cells, an efficient mechanism of transport of iodothyronines across the plasma membrane is required. Recently, it has been suggested that also complexes hormone-binding proteins may enter into the cell, thus providing another pathway for cellular uptake of biologically active hormones (Hammes et al. 2005).

Up to now, several TH transporters have been identified, such as the Na⁺/taurocholate cotransporting polypeptide (SLC10A1), multidrug resistance-associated proteins, the heterodimeric L-type amino acid transporters (LAT1 and LAT2), the organic anion-transporting polypeptide (OATP) family, and the MCTs family. Most of these transporters are not specific for T3, since they bind to different ligands, with the exception for MCT8 (SLC16A2), MCT10 (SLC16A10), and OATP1C1 (SLCO1C1) (Visser et al. 2011).

T3 concentrations in the CNS are roughly the 20 % of serum levels (Schroeder and Privalsky 2014). In order to reach the brain, the TH need to cross the blood–brain barrier (BBB). This is a selective permeability barrier, composed of endothelial cells of brain capillaries surrounded by the astrocyte processes, which separates the blood from the brain extracellular fluid.

TH enter in neurons by two mechanisms, mediated by OATP1C1 and MTC8, respectively. In the first case, TH uptake occurs from the endothelial cells into the astrocytes, with a greater efficiency for T4 than T3. Thereafter DIO2 converts it locally to T3, which enters in the neurons by binding to MCT8. Alternatively, TH may also enter directly in the neurons through gaps in the astrocytes processes via MTC8, which exhibits a higher affinity for T3 compared to T4 (Mayerl et al. 2012).

MCT8 is also expressed in the heart, liver, kidney, adrenal glands, and thyroid, but its function appears to be critical especially in the brain: in fact genetic defects of this transporter are involved in the pathogenesis of the Allan–Herndon–Dudley syndrome and of the Pelizaeus–Merzbacher-like disease, which are associated with a severe neurological phenotype. Data about the role of MCT10 and OATP1C1 in human physiology are limited (Visser et al. 2011).

12.1.3 Nuclear Receptors

Thyroid hormone receptors belong to the nuclear receptor superfamily. They bind as heterodimers with retinoid X receptor (RXR), or less frequently as homodimers, to regulatory DNA sequences, known as thyroid response elements (TRE) located in the promoter of target genes. A number of cofactors (proteins acting as coactivators and corepressors) are involved in TH receptor signaling.

On genes positively regulated by TH, in the absence of T3, the hetero/homodimers associated to corepressors, bind to TREs and repress transcription. Binding of T3 to TRs results in dissociation of corepressors, recruitment of coactivators, and transcriptional activation.

In contrast, negatively regulated TH target genes show transcriptional activation in the absence of TH and repression in the presence of the ligand T3.

The two major corepressors, the nuclear receptor corepressor (NCoR) and the silencing mediator of retinoic acid and thyroid hormone receptors (SMRT), are crucial regulators of nuclear receptor signaling (Astapova et al. 2008). They form a complex with other repressors, such as Sin 3, and histone deacetylases (Hu and Lazar 2000), thus local chromatin structure seems crucial to shuts down basal transcription. Several coactivators interact with TRs, such as the steroid receptor coactivator complex (SRC) and the vitamin D receptor interacting protein–TR associated protein complex (DRIP–TRAP), which enhance the T3-dependent transcription. SRC complex interacts with CREB-binding protein (CBP), responsible for cAMP-stimulated transcription, interacting with the phosphorylated form of CREB (cAMP-regulated enhancer binding protein) and with the related protein p300. CBP/P300 interact with P/CAF (p300/CBP-associated factor) which has an intrinsic histone acetyltransferase (HAT) activity and with the RNA pol II (Torchia et al. 1998). DRIP–TRAP complex, which is homolog of the yeast Mediator complex, does not appear to have intrinsic HAT activity, however several components of this complex associates with RNA Pol II, thus connecting nuclear receptors to the basal transcriptional machinery (Ito and Roeder 2001).

Two receptors (TRα and TRβ) mediate the TH effects at the nuclear level. They are encoded by two separate genes, located on human chromosomes 17 and 3, respectively. Two TRβ isoforms have been identified: TRβ2 is mainly expressed in the hypothalamus, pituitary, retina, and inner ear; conversely, TRβ1 is the principal isoform in liver and kidney. The TRα1 predominates in the CNS, skeletal, intestine, and cardiac muscle. The TRα2 isoform differs for the TRα1 in the C-terminus and is unable to bind T3, but retains DNA-binding properties and it seems to have a weak antagonistic effect.

With regard to central nervous system (CNS), the TR-α1 is the isoform that accounts for about 80 % of the TH receptors expressed in brain. Compared to TR-α, the TRβ is expressed at a later stage of brain development. TRβ1 isoform is widely expressed in the brain (Bradley et al. 1992), whereas the TRβ2 isoform is mainly expressed in the hypothalamus and pituitary gland (Hodin et al. 1989; Lechan et al. 1994). In addition, the TRβ1 and TRβ2 isoforms are expressed in a neurogenic subpopulation, located in hippocampus and adult brain which seems involved in the proliferation of neuronal progenitors (Desouza et al. 2005). In particular, it has been suggested that unliganded TR-β isoform may exert an inhibitory effect on hippocampal cellular growth, as suggested by the reduced proliferation of these neurons in hypothyroid mice and the increased proliferation associated with deletion of both TR-β1 and TR-β2, in mice (Kapoor et al. 2011). Similarly patients with THRB homozygous deletion do not display an intellectual impairment (Refetoff et al. 1967). An analogous behavior has been hypothesized for unliganded TR-α1 in the cerebellum (Fauquier et al. 2014; Heuer and Mason 2003) and hippocampus (Kapoor et al. 2010).

12.2 Resistance to Thyroid Hormones Syndrome Due to Mutations in THRB Gene (RTH β)

12.2.1 General Clinical Features

Resistance to thyroid hormones syndrome (RTH β) is a rare condition and more than 3000 cases have been published from about 1200 different families with a wide geographic and ethnic distribution (Refetoff et al. 1993; Gurnell et al. 2010; Dumitrescu and Refetoff 2012). The actual prevalence of the disease is unknown, since the routine screening programs for congenital hypothyroidism are based on the TSH measurement and consequently cannot diagnose this condition. A limited survey in a cohort of 80,000 newborns found one case per 40,000 live births.

The majority of the cases (nearly 85 % of the case) are associated with heterozygous mutations in the TRβ gene and the condition is inherited in an autosomal dominant fashion (Refetoff et al. 1967).

This inheritance depends on the dominant negative effect, due to the inhibition of the activity of the wild-type β- and α-receptors, by the mutant TR-beta. These mutant receptors display either a reduced affinity for T_3 or an impaired interaction with the cofactors (coactivators and corepressors), thus losing its ability to modulate target gene expression in different tissues.

Different mechanism have been evoked to explain this dominant negative effect (Dumitrescu and Refetoff 2012):

1. The formation of inactive dimers between mutant TRs and wild-type TRs.
2. A competition between mutant and wt receptors for essential cofactors.
3. A competition between mutant TR and wild-type TR for DNA-binding sites.

In the original RTH β family, in which a deletion of exons 4-10 resulted in the abolition of the dimerization and DNA-binding properties of TRβ, the disease segregated as an autosomal recessive trait. The homozygous patients had goiter and deaf-mutism together with high TH levels; conversely, the heterozygous subjects were phenotypically normal, supporting the hypothesis that reduced amount of TRβ does not produce haploinsufficiency (Refetoff et al. 1967; Takeda et al. 1992) and that the mutant receptor must conserve its DNA-binding and dimerization properties, in order to cause the phenotype of RTH β.

The TRβ mutations are distributed in the carboxyl terminus of the TRβ. Typically, three CpG-rich "hot spots" regions are located in the ligand-binding domain and in the contiguous hinge domain of the protein.

In contrast to what is observed for other nuclear receptors (such as vitamin D, androgen receptor or PPARγ), no mutations have been identified in the DNA-binding domain or in other regions of the receptor. It is likely that heterozygous mutations in these regions may be clinically silent.

In about 10–15 % of the cases with clinical and biochemical phenotype of RTH β, no mutation could be found in the TRβ gene and this situation is defined as "non-TR–RTH." It is speculated that these patients may have an abnormality of one of the cofactors or TH transporters into the cells. However screening of several families with non-TR-RTH excluded the involvement of coactivators (SRC-1/NcoA-1; and NcoA-3/SRC-3/AIB1/RAC-3), two corepressors (NCoR and SMRT) and two coregulators (RXRγ and TRIP1) as well as the cell transporter LST-1 (OATP1B1) (Reutrakul et al. 2000).

There are no clinical signs or symptoms typical of RTH β: the clinical picture is wide ranging from asymptomatic cases to subjects who manifest symptoms of thyrotoxicosis.

Differences in the degree of hormonal resistance are linked to the different levels of TRβ and TRα expression, in different tissues.

TRβ mainly is expressed in the hypothalamus, kidney, liver, anterior pituitary gland, hypothalamus, retina, and cochlea, whereas TRα predominates in the skeletal and cardiac muscle, brain, brown fat, intestine, spleen, and vascular endothelial cells. Consequently, symptoms of TH deficiency and excess could coexist. As an example, hypercholesterolemia, delayed bone maturation, growth retardation, and learning disabilities are suggestive of hypothyroidism, while weight loss, heat intolerance, hyperactivity, and tachycardia are compatible with thyrotoxicosis.

Classically, RTH β subjects have been classified into two subgroups according to the absence or presence of symptoms of thyrotoxicosis, selective pituitary resistance (PRTH), and generalized thyroid hormone resistance (GRTH), respectively. Patients with PRTH display variable symptoms of hyperthyroidism (Beck-Peccoz and Chatterjee 1994; Beck-Peccoz et al. 2006). Conversely, subjects with GRTH exhibit a sort of "compensated hypothyroidism," being the genetic defect of TH responsiveness balanced by the high circulating TH concentrations; the efficiency of this compensatory mechanism is variable in each individual, in different tissues, as well as in different periods of life.

In addition, TRβ mutations found in both GRTH and PRTH may be the same and patients of the same family may present with either form. Indeed, PRTH patients have normal levels of sex-hormone-binding globulin, a marker of peripheral thyroid hormone action, elevated in the case of hyperthyroidism, thus suggesting that insensitivity to TH action is present not only in the hypothalamic–pituitary region, but also in the liver (Beck-Peccoz et al. 2006). However such distinction may be clinically helpful.

The main clinical features of patients with RTH β are summarized in the following paragraphs.

Goiter

Diffuse or multinodular goiter is a common finding in RTH β, independently from the presence of clinical symptoms. An increased biological activity of circulating TSH molecules may be involved in the pathogenesis of goiter in RTH β subjects, who have normal TSH levels (Gurnell et al. 2010). In RTH β patients treated with surgical ablation, the goiter commonly relapse with nodular alterations and gross asymmetries, requiring additional surgery or radioiodine.

Cardiovascular Symptoms

Approximately 75 % of RTH β patients exhibits palpitations and tachycardia at rest. Predominance of TRα may explain the presence of partially hyperthyroid response in the heart, as the few mutated TRβ shall exert less dominant negative effect on the normal receptors. The finding that some indices of cardiac systolic and diastolic function (e.g., heart rate, stroke volume, cardiac output, diastolic filling, maximal aortic flow velocity) showed values that are intermediate between normal and hyperthyroid subjects supports this hypothesis. However other parameters (e.g., ejection and shortening fractions of the left ventricle, systolic diameter, and left ventricle wall thickness) were not different, suggesting an incomplete response of the heart to the high TH concentrations. In addition, systemic vascular resistance and arterial stiffness are increased in RTH β, as seen in subclinical hypothyroidism, thus indicating a more complex derangement of cardiovascular function. One study suggested an increased incidence of mitral valve prolapse among RTH β subjects. Finally, a reduced whole-body insulin sensitivity and dyslipidemia have been documented in a number of patients, suggesting an increased cardiovascular risk in RTH β (Kahaly et al. 2002; Pulcrano et al. 2009; Owen et al. 2009).

Skeletal Abnormalities

Studies performed in animal models with a PV mutation targeted to the TRβ gene suggest that skeletal thyrotoxicosis, due to elevated circulating thyroid hormone levels which overstimulate the intact TRα1 signaling pathway, may be responsible for bone abnormalities in RTH β (O'Shea et al. 2006).

In humans, dysmorphic skeletal features, such as "stippled epiphyses," dysmorphic facies, and winged scapulae, have been documented only in the cases harboring homozygous deletion of TRβ gene. Dimorphic facies (birdlike facies with prominent nasal bridge) are also associated with homozygous THRB mutations.

Delayed bone maturation and growth are present in about one-third of children with RTH β.

Although height below the fifth percentile for age and sex is a relatively common finding during childhood, the final adult height seems not different from the unaffected relatives.

A decreased bone mineral, which may cause precocious osteoporosis and increasing risk of fractures, have been reported in adult RTH β. Conversely, the normal levels of the markers of bone turnover may imply a reduced bone formation rate resulting in a low peak bone mass. These findings are similar to what is observed in childhood hypothyroidism.

Metabolism

Low body mass index (BMI) is reported in about 30 % of RTH β children, in spite of the hyperphagia and the enhanced energy intake.

Basal metabolic rate (BMR) has been found normal or increased, particularly in PRTH patients. Indirect calorimetry assessment showed enhanced resting energy expenditure (REE), either in adults or in children with TRβ mutations. This increase was intermediate between euthyroid and thyrotoxic subjects. Skeletal muscle and myocardium, in which the TRα isoform expression is prevalent, seem responsible of increased energy expenditure, as suggested by the correlation between mean heart rate and REE in both RTH and thyrotoxicosis. In both these conditions, TH excess was associated with uncoupling between tricarboxylic acid cycle activity and ATP synthesis in vivo, as measured by magnetic resonance spectroscopy (Mitchell et al. 2010).

Immune System

An increased frequency of respiratory infections (pulmonitis and infections of the upper respiratory tract) has been reported in RTH β patients, compared to their unaffected relatives. This susceptibility has been related to reduce immunoglobulin concentrations. In addition receptors for TH are present in granulocytes and lymphocytes.

In mothers affected with RTH β, there is a higher rate of miscarriage and intrauterine growth retardation of unaffected offspring, thus suggesting that intrauterine exposure to high TH levels does have adverse effects on the fetus.

Other Features

There is only one patient, homozygous for TRβ mutation, in whom RTH β may have contributed to death: this patient had resting pulse of 190 beats/min and died from cardiogenic shock complicated by septicemia.

Coexistence of TSH-secreting pituitary adenomas (TSHomas) and RTH β has been suggested in only two cases. The impaired TH feedback in the pituitary may lead to a continuous stimulus to thyrotropes to synthesize and secrete TSH molecules, which may play a role in the development of pituitary tumors. However, the pituitary lesions associated to RTH β appear to be pituitary "incidentalomas" (Beck-Peccoz and Persani 2010). Interestingly, somatic mutation of TR-beta has been found in two TSH secreting pituitary adenomas (Ando et al. 2001a, b)

Occasionally, RTH β occurs in association with autoimmune thyroid disorders, such as Graves' disease or Hashimoto's thyroiditis. The occurrence of anti-TPO or anti-TSH receptor autoantibodies in RTH subjects has been described. Recent data suggest that the individuals with RTH β due to TRβ gene mutations have an increased likelihood of AITD compared to unaffected relatives (Barkoff et al. 2010). The reason of this association seems related with the hyper stimulation, via TR-alpha, of the cells of the immune system.

The RTH patients, who develop Graves' disease, undergo a progressive increase in goiter size along with frank symptoms of thyrotoxicosis. The further elevation of TH levels causes TSH secretion to be totally inhibited. Conversely, hypothyroidism may occur in the presence of normal serum TH concentrations, as a consequence of Hashimoto's thyroiditis.

12.2.2   Differential Diagnosis of RTH

The differential diagnosis with TSH-secreting pituitary adenomas, which is characterized by a similar biochemical features of elevated serum concentrations of both FT4 and FT3, associated with a normal or slightly elevated serum TSH, is mandatory (Agrawal et al. 2008). The presence of the same biochemical pattern of thyroid function in other first-degree relatives supports the diagnosis of RTH β, since familiar cases of TSH-oma have never been reported (except for four families in a setting of Multiple Endocrine Neoplasia 1). In these cases, molecular analysis of the THRB gene makes a definitive diagnosis in 85–90 % of cases of RTH β.

Although different clinical parameters have been proposed (basal metabolic rate, systolic time intervals, Achilles reflex time) in order to discriminate among these two conditions, the clinical presentation of patients with RTH β, particularly of PRTH, and those with TSH-oma (Beck-Peccoz and Persani 2010) is similar.

In patients with TSH-omas, serum levels of glycoprotein hormone α-subunit (α-GSU) and α-GSU/TSH molar ratio, corrected for age, sex, and circulating levels of gonadotropins, are elevated, whereas in RTH β patients both indices are in the normal range.

To assess the degree of resistance in specific target tissues, different in vitro parameters have been proposed such as sex hormone-binding globulin (SHBG), angiotensin converting enzyme (ACE), carboxyterminal telopeptide cross-linked of type 1 collagen (ICTP), soluble interleukin-2 receptor (sIL-2R), osteocalcin, cholesterol, creatinine kinase, and ferritin. Particularly, SHBG and ICTP are clearly elevated in patients with TSH-oma, compared with RTH. The sensitivity and specificity of these tests is improved, when assessed after T3 suppression test, performed with oral administration of supra-physiological doses of T3 (50 µg/day for 3 days, followed by 100 µg/day for other 3 days and then 200 µg/day for other 3 days) (Cheng et al. 2010). In RTH β patients, the increase of peripheral markers of TH actions and heart rate is blunted in comparison to normal subjects, thus definitively confirming the presence of resistance to TH action.

As far as the dynamic testing is concerned, TRH test (iv injection of TRH 200 µg) has been widely used: in the majority of patients affected with TSH-oma, TSH and α-GSU levels do not increase after TRH injection, whereas RTH β subjects show normal response of TSH. An exaggerated response is found in patients who underwent thyroid ablation or were treated with antithyroid drugs.

T3 inhibitory test, performed as reported above or administering T3 for 8–10 days at the dose of 80–100 µg/day, may show a full inhibition of TSH levels in RTH β patients, but persistent TSH response to TRH, carried out at the end of T3 administration. Since none of these tests have a clear diagnostic cut-off value, the combination of them, if possible, increases the specificity and sensitivity of the diagnostic process.

Pituitary MRI is required in case of not univocal results with other tests, however the detection of pituitary lesion does not definitely rule out the diagnosis of RTH β. In fact, pituitary lesions are a quite common finding (20–25 % of MRI performed for other reasons) in the general population. These lesions are usually considered as "pituitary incidentalomas," especially when a hypothalamic–pituitary dysfunction has been excluded. The presence of a microadenoma, in combination with lack of TSH response to dynamic tests and high levels of α-GSU or α-GSU/TSH molar ratio strongly sustains the diagnosis of a TSH-oma.

12.2.3 Neurological and Cognitive Impairment in Resistance to Thyroid Hormone

It has been hypothesized that in RTH β an uncompensated hypothyroidism at an early stage may be responsible for defects of neuroanatomical development.

Few data are available about the brain anatomical abnormalities associated with RTH β. A single MRI study in 43 RTH β patients found, in male patients, an increased frequency of cerebral anomalies of the left hemisphere, particularly an extra or missing gyri in the parietal bank of the Sylvian fissure or multiple Heschl's transverse gyri in the primary auditory cortex when compared to unaffected relatives. No patent abnormalities were found in female patients (Leonard et al. 1995).

Although severe mental retardation (IQ <60) is uncommon (only 3 %), about 30 % of affected subjects display a mild learning disability (IQ <85). In particular either the verbal or the performance component was impaired compared with controls (Brucker-Davis et al. 1995). Some authors have reported in their RTH β cohort a high frequency of attention deficit hyperactivity disorder (ADHD). This finding has not been confirmed by other groups, but it is possible that the low IQ may be responsible for ADHD manifestations, more than RTH β per se. In addition, an increased frequency of delayed developmental milestones and language disorders have been found in RTH β patients, compared to their unaffected relatives (Brucker-Davis et al. 1995; Stein et al. 1995; Hauser et al. 1993; Weiss et al. 1997).

The neuroanatomical regions involved in attention and vigilance are located in the right lateral prefrontal cortex, in the parietal lobe, and in anterior cingulate. Consistently, Matochik et al. found a severe impairment on an attention auditory discrimination task, in adults with RTH β compared to controls. The PET scan performed during this task demonstrated the presence of an increased metabolic activation of the anterior cingulate in RTH β. The reduction of the functional activity in this brain area and the subsequent activation of other structures, such as the frontal cortex, are required for an efficient performance on complex attention tasks. However, it is not clear whether these functional anomalies are related to a defect in brain development or may be a consequence of the elevated levels of thyroid hormones via overstimulation of the TR-α (Matochik et al. 1996).

Genotype–Phenotype Correlation of Cognitive Abnormalities in RTH β

Patients with homozygous deletion of THRB display a phenotype characterized by deaf mutism due to sensorineural hearing loss, delayed bone maturation, stippled epiphyses, goiter, and high levels of circulating thyroid hormone in the presence of a normal TSH (Refetoff et al. 1967; Takeda et al. 1992).

Interestingly, these patients do not display growth delay, mental retardation, or cognitive impairment, while the five cases, homozygous for THRB mutations, are invariably associated with an intellectual impairment. The common features of these subjects are the extreme symptoms of resistance, the hyperactivity and the hearing loss, associated with a variable degree of mental retardation.

In particular, the first patient described had a homozygous deletion of threonine 337 in TR-β. He presented with respiratory distress, hyperbilirubinemia, goiter, tachycardia, severe mental retardation, and seizures. This child died at 8 years as a consequence of staphylococcal pneumonia and cardiogenic shock (Ono et al. 1991).

A second patient with a homozygous/hemizygous mutation in the THRB gene (I280S) showed hyperactivity and a severe cognitive impairment (IQ below 60) with no active speech; in addition he suffered for hearing loss and he was unable to walk. The severity of the clinical picture had been probably worsened by the administration of thionamides during the first months of life (Frank-Raue et al. 2004).

Recently, other three homozygous patients have been described by Refetoff et al. Similarly to the previous cases, these patients displayed severe symptoms of resistance to thyroid hormone. In one case a homozygous single nucleotide change in the THRB gene resulted in the substitution of the glycine 347 with a glutamic acid (G347E). This patient presented with a delayed verbal development and a reduced IQ score. In another family, patients homozygous for the R316C mutation showed a mild mental retardation in one case and a moderate mental retardation in the other sibling. A common neurological feature of these homozygous patients is the hyperactivity and the hearing loss (Ferrara et al. 2012).

"Conventional" heterozygous mutations, resulting in a premature stop codon with the consequent production of a TR-β lacking a number of residues in the C-terminal, display a strong dominant-negative effect in vitro and are often associated with a more severe clinical phenotype, including mental retardation. A girl harboring a single nucleotide change, resulting in the replacement of cysteine codon 434 by a stop codon with the deletion of the last 28 amino acids of the wild-type protein, had an extremely retarded mental and physical development and she had to attend a special school for mentally hindered children (Behr et al. 1997).

A second case was heterozygous for a frameshift mutation, causing the production of a truncated receptor lacking the last 20 amino acids. The affected girl had goiter, growth retardation, short stature, and deafness associated with hypotonia, mental retardation, visual impairment, and seizures (Phillips et al. 2001).

However, the demonstration of a precise genotype–phenotype correlation is lacking. As an example, in a single patient the E449X mutation was associated with a severe neuropsychomotor retardation associated with irritability and aggressiveness, not observed in a previous case with the same mutation (Gurgel et al. 2008).

Visual System

In animal models, deletion of the TRβ2 isoform produces a selective loss of M-cone photoreceptors resulting in abnormal color vision. In particular, during embryogenesis, TR-β seems responsible for the photoreceptor distribution in the retina, inhibiting the S-opsin and committing the differentiation of M opsin photoreceptor. However, no abnormalities of color sensitiveness have been identified in "conventional" RTH β patients with heterozygous TRβ mutations. Patients with homozygous deletion of THRB gene (Refetoff et al. 1967; Takeda et al. 1992) are color blind, while in one patient with compound heterozygous mutation (R338W in exon 9 and R429W in exon 10 of THRB gene), an abnormal electroretinographic pattern was found, characterized by a normal scotopic response and a reduced photopic response. In particular, this patient showed a small amplitude b-wave to a red flash and a larger amplitude b-wave to the blue flash, similar to what is commonly described in the enhanced S cone syndrome (Weiss et al. 2012).

Hearing System

An increased incidence of conductive or sensorineural hearing impairment, which may contribute to the defective speech development, has been reported in some RTH β children. The patho-physiology of these abnormalities is composite, being the conductive defect due to the higher susceptibility to upper airways infection of RTH β children, whereas the defective TRβ expression may be responsible for the cochlear dysfunction (Brucker-Davis et al. 1996).

Noteworthy, mice with targeted disruption of the TRβ locus develop profound sensorineural hearing loss, thus suggesting an important role of TH in the development of the hearing system.

12.2.4 *Therapy*

There is currently no definite therapy to correct the molecular defect causing RTH β; however, in most patients a specific treatment is not necessary. In this case, goiter may be the only sign of the disease. In fact, these subjects are clinically euthyroid being high levels of circulating free TH able to compensate for the resistance in peripheral tissues.

Patients with signs and symptoms of thyrotoxicosis, such as tachycardia and palpitations at rest, may benefit by the use of a cardioselective β-blockers (atenolol or others) not inhibiting the peripheral conversion of T4 to T3, thus worsening a possible hypothyroid state in some tissue. In the event of severe thyrotoxic symptoms, not responding to β-blockers, a reduction of thyroid hormone levels may be beneficial. This can't be achieved using antithyroid drugs, because the consequent increase of TSH levels may determine goiter enlargement and pituitary hyperplasia.

The treatment of choice in such cases is the administration of thyromimetic compounds, such as 3,5,3′-triiodothyroacetic acid (TRIAC), which through the feedback mechanism, reduces TSH secretion and causes a slight decrease of circulating T4 levels (values of T3 are unreliable as TRIAC cross react in T3 measurement methods). TRIAC has a weaker effect on peripheral tissues, compared to T3. As a consequence the administration of this drug reduces the thyrotoxic signs and symptoms, particularly at the heart level.

It has been shown to be beneficial in both children and adult patients with RTH β at the dose of 1.4/2.8 mg/day fractionated in twice or thrice daily administrations (Refetoff and Dumitrescu 2007; Gurnell et al. 2010; Beck-Peccoz et al. 1983).

The use of dopaminergic drugs and somatostatin analogs has limited success because TSH secretion rapidly escapes the inhibitory effects of both drugs, as the T4 reduction triggers the much more potent stimulatory effect of TH negative feedback mechanism.

Although controversial, in children with signs of growth or mental retardation, the administration of supraphysiological doses of L-T4 to overcome the high degree of resistance present in some tissues can be beneficial. Supraphysiological doses of thyroid hormones are also necessary in patients treated with total thyroidectomy for a missed diagnosis of RTH β. The use of high doses of L-T4 requires a careful monitoring of patients, assessing the indices of peripheral thyroid hormones action.

Recently, TRβ selective agonists (GC1, eprotirome) have been developed and they could be beneficial for some abnormalities (dyslipidemia) found in RTH β. Unfortunately, the development program on this drugs has been discontinued after the evidence of cartilage damage after 12 months administration in dogs. In addition there is evidence that eprotirome may induce liver injury in humans (Sjouke et al. 2014).

Conversely, the development of TRα selective antagonists that may also be useful in controlling symptoms of thyrotoxicosis mediated through the TRα.Therapy It is conceivable that future treatments shall be based on administration of small molecules that block TSH receptor activity on thyroid cell surface (Neumann et al. 2011). Moreover, the possible application of gene therapy to RTH β patients will aim to either silence the mutated receptor or repair the mutated nucleotide responsible for the expression of abnormal receptor protein.

12.3 Resistance to Thyroid Hormones Due to THRA Mutations

12.3.1 General Features

Recently, the first three families with TH resistance due to TR-alpha (RTH α) have been described (Bochukova et al. 2012; van Mullem et al. 2012; Schoenmakers et al. 2013; Moran et al. 2013). Similar to what is described in animal models (O'Shea et al. 2005; Dumitrescu et al. 2005), these subjects present variable features of hypothyroidism, associated with near-normal TH and TSH levels.

The clinical presentation of RTH α is characterized by abnormalities in tissues in which the TRα is the major isoform expressed, thus growth and developmental retardation, skeletal dysplasia, constipation, and a mild cognitive defect seem the main features of this syndrome.

Free T4 levels were slightly below the lower end of the normal range while free T3 levels at the upper end of normal, resulting in a reduced FT4/FT3 ratio; circulating TSH was normal.

Interestingly the four affected individuals of the three families showed a truncated form of the receptor (E403X, F397fs406X and Ala382ProfsX7) with a premature stop codon located in exon 9, thus affecting the only TR alpha 1 isoform and not the other transcript (TRα2, Rev-erbα) generated from the THRA locus. All these mutations showed in vitro a reduced transcriptional activity and a strong dominant negative effect on wt receptor.

Common features of this syndrome are growth retardation, which transiently improves after L-T4 administration, disproportionate short stature, macrocephaly, low free T4/free T3 ratio and rT3 levels, together with subnormal heart and basal metabolic rate.

Delayed tooth eruption, delayed closure of skull sutures, and femoral epiphyseal dysgenesis are also present; consistently mice with TRα1-PV mutation showed reduced endochondral and intramembranous ossification, postnatal growth retardation, and delayed closure of skull sutures (O'Shea et al. 2005; Dumitrescu et al. 2005).

Constipation due to delayed intestinal transit is present in all the four cases described up to now; however in the family with the TRα1-F397fs406X, the administration of L-T4 improved this symptom in both father and affected daughter, having stools every day when receiving therapy (Schoenmakers et al. 2013).

Interestingly in all the affected patients low or low-normal levels of IGF-1 were found. One patient was treated with hr-GH, without a significant improvement of the growth retardation (Schoenmakers et al. 2013). Also the L-T4 administration was only transiently beneficial on the growth delay.

In summary, these patients retain a normal hormone responsiveness in the hypothalamic–pituitary axis and liver but they display a skeletal, gastrointestinal, and myocardial resistance (Table 12.2).

12.3.2 Neurological and Cognitive Impairment in RTH α

In the first pediatric case described, the cognitive deficits were consistent with a congenital hypothyroidism (van Mullem et al. 2012). She was inappropriately placid, her speech was slow and monotonous. A neuropsychological assessment showed selective cognitive deficits in the adaptive behavior, in the short-term memory, and in the visuoperceptual function, while the verbal comprehension was in the normal range. In addition she experienced motor dyspraxia associated with difficulties in fine motor coordination resulting in the inability to write or draw. In addition she had a broad-based ataxic gait. Finally muscular hypotonia, but not weakness, was present.

A similar phenotype, thus associated with a more severe cognitive impairment, has been described in another female patient harboring the Ala382ProfsX7 mutation. The patient was unable to read and her IQ was around 52. In addition, this patient was affected with epilepsy, confirmed by the electroencephalographic demonstra-

Table 12.2 The clinical phenotype of the first four cases affected with RTH α

Phenotype	TRα1-E403X	TRα1-F397fs406X	Ala382ProfsX7
Femoral epiphyseal dysgenesis	+	+	+
Delayed tooth eruption	+	+	NA
Delayed closure of skull sutures	+	+	NA
Growth retardation	+	+	+
Delayed bone age	+	+	+/−
Macrocephaly	+	+	+
Low T4/T3 ratio	+	+	+
Low rT3	+	+	+
Low IGF1	+	+/−	+/−
Raised SHBG	+	−	−
Reduced GI motility	++	+	+
Motor abnormality	+	+	+
Seizures	−	−	+
Cognitive impairment	+	+	+
Reduced basal metabolic rate	+	NA	+
Reduced resting heart rate and blood pressure	+	NA	+

Data from Schoenmakers at al. (2010)

+ Present, ++ severe, +/− low/normal, *NA* not assessed

tion of bilateral theta waves during hyperventilation; seizures decreased in frequency with sodium valproate administration (Moran et al. 2013).

The proband of the second family (TRα1-F397fs406X) and her affected father had a mild cognitive deficit with an IQ of 90 and 85, respectively (Schoenmakers et al. 2013).

The observed neurocognitive deficits seem associated with structural abnormalities such as microcefaly, reduced cerebellar and hippocampal volume, diminished white matter density and accord with known developmental actions of TH and substantiate the critical role of TRα1 in CNS (Moran C, 2014 BES meeting, personal communication).

Although the third patient described had hypermetropia, the visual system and the hearing system does not seem affected by the THRA mutation.

12.3.3 Therapy

The administration of L-T4 resulted in normalization of FT4, with a further increase of FT3 and suppressed TSH levels, suggesting a conserved negative feedback of TH on TSH secretion.

In patient with E403X mutant, the levels of IGF-1 normalized, without a real improvement in growth velocity, growth rate and intestinal transit time did not change significantly. Heart rate and blood pressure did not improve. In the second family, L-T4 treatment caused an improvement of constipation with a persistence of growth retardation in both subjects.

Higher-dose thyroxine therapy or the use of TRα-selective thyromimetic agents may be necessary to avoid hyperthyroidism in TRβ-expressing tissues (Schoenmakers et al. 2013).

12.4 Disorders of Thyroid Hormone Metabolism

12.4.1 General Features

Iodothyronine deiodinases (DIOs) are a family of selenocysteine containing enzymes, required for activation or inactivation of thyroid hormones and their metabolites.

Although the most common alterations of TH metabolism are acquired, such as the "low T3 syndrome" of non-thyroidal illness, genetic conditions associated with defective function of deiodinases have been recently described. Mutations in SECIS-binding protein 2 (SBP2), a key protein that allows the incorporation of selenium in selenoproteins, cause defective production of DIOs. Being the seleno-proteins ubiquitous and multifunctional, those individuals manifest a complex phenotype alongside the abnormal thyroid function (Refetoff and Dumitrescu 2007; Agrawal et al. 2008; Schoenmakers et al. 2010).

The main laboratory finding is an abnormal pattern of thyroid function characterized by high free T4, low free T3, raised reverse T3, associated with normal or slightly elevated TSH levels.

Up to now only six families exhibiting reduced TH sensitivity due to a disorder of thyroid hormone metabolism have been described. The defect is inherited in an autosomal recessive fashion and it is caused by homozygous and compound heterozygous mutations in the SBP2 gene.

This defect is caused by mutations in SBP2, a factor required for the incorporation of selenocysteine during selenoprotein biosynthesis. Selenoproteins are a family of about 25 proteins with wide functions, which include metabolism of thyroid hormones (deiodinases), removal of cellular reactive oxygen species, reduction of oxidized methionines in proteins, and transport and delivery of selenium to peripheral tissues. The rare amino acid, selenocysteine (Sec) is critical for their enzymatic activity.

A multiprotein complex, including SBP2, is responsible for the incorporation of selenium into selenoprotein. A specific stem-loop (called SECIS elements) in the 3′-UTR region of selenoprotein mRNAs interacts with and leads to seleno-cysteine incorporation at UGA codons (Refetoff and Dumitrescu 2010). Defects in this machinery result in miscoding of the UGA as a signal to stop synthesis and the transcript may undergo decay. The affinity of SBP2 for SECIS elements of different mRNA is variable and this contributes to hierarchy in selenoprotein production in case of defective function of SBP2 or Se deficiency. In mice complete disruption of SBP2 is embrionically lethal. In human the SBP2 mutations described up to now cause a severe reduction of selenoprotein, but not a total depletion. This is probably due to the highly complex architecture of SBP2, with internal methionine residues capable of starting the synthesis of shorter protein isoforms.

Beside the abnormalities in TH metabolism, the phenotype of affected individuals is highly variable ranging from milder to more severely affected individuals.

Deficiencies of multiple selenoproteins have been documented in all cases: glutathione peroxidases are markedly reduced, and circulating levels of hepatic selenoprotein P are low, accounting for the low serum selenium levels recorded in these families. Childhood growth retardation is a common feature in all the families described (Schoenmakers et al. 2010; Di Cosmo et al. 2009; Azevedo et al. 2010; Hamajima et al. 2012). The only adult subject described was azoospermic, with reduced levels of testis-enriched selenoproteins that causes spermatogenic arrest. In addition, he was markedly photosensitive, with a dermal deficiency of antioxidant selenoenzymes causing increased cellular reactive oxygen species, membrane lipid peroxidation, and oxidative DNA damage. Reduction of antioxidant enzymes in immune cells was associated with impaired T cell proliferation and shortened telomeric DNA. The latter was associated with anemia and lymphopenia, similar to what is observed in aplastic anemia found in telomerase deficiency. Increased adipose mass and increased insulin sensitivity have been described in two families and eosinophilic colitis was found in one individual.

12.4.2 Neurological and Cognitive Impairment in Disorders of Thyroid Hormone Metabolism

The first three families described did not display a specific neurological phenotype (Dumitrescu et al. 2005; Di Cosmo et al. 2009), while the affected individuals recently described showed a more severe impairment. A 12-year girl presented with hypotonia and weakness early in her life (Azevedo et al. 2010). She had motor coordination disorder with delayed motor and intellectual milestones (walking at age 2 years and speaking at 3 years); later in childhood she developed a symmetrical peripheral sensitive neuropathy confirmed by electroneuromyography and somatosensory evoked potential test, characterized by a slow progression. Brain RMI was normal while the audiometric test showed bilateral sensorineural loss. At the age of 11 years she was mentally retarded and suffered for a progressive peripheral myopathy, similarly to what is observed in patients with mutations in the SEPN1 gene.

Mild bilateral high-frequency hearing loss was also found in a male child presented age 2 years with failure to thrive, developmental delay and short stature and in an adult patient presented at 35 years for infertility and fatigue. Similarly to what is observed in the previous family, they both showed delayed motor and speech development and myopathy resulting in muscle weakness. In two patients an increased frequency of exudative otitis media and rotary vertigo was found (Schoenmakers et al. 2010; Azevedo et al. 2010). A 10-year-old Japanese boy and an 11-year-old Turkish girl born showed a mild mental retardation (Dumitrescu and Refetoff 2012; Hamajima et al. 2012).

12.4.3 Therapy

Clinical trials with oral selenium supplementation showed a raised circulating selenium concentrations, without improving the thyroid abnormalities (Di Cosmo et al. 2009; Schomburg et al. 2009). T3 treatment was clearly beneficial for growth in two children. Tocopherols, lycopene, and other antioxidant agent may be beneficial in reducing the oxidative damage, as suggested by preliminary in vitro experiments. In one Japanese case treatment with rhGH combined with T3 improved both longitudinal bone growth and maturation (Hamajima et al. 2012).

References

Agrawal NK, Goyal R, Rastogi A et al (2008) Thyroid hormone resistance. Postgrad Med J 84:473–477

Ando S, Sarlis NJ, Krishnan J et al (2001a) Aberrant alternative splicing of thyroid hormone receptor in a TSH-secreting pituitary tumor is a mechanism for hormone resistance. Mol Endocrinol 15:1529–1538

Ando S, Sarlis NJ, Oldfield EH et al (2001b) Somatic mutation of TRbeta can cause a defect in negative regulation of TSH in a TSH-secreting pituitary tumor. J Clin Endocrinol Metab 86:5572–5576

Astapova I, Lee LJ, Morales C et al (2008) The nuclear corepressor, NCoR, regulates thyroid hormone action in vivo. Proc Natl Acad Sci U S A 105:19544–19549

Azevedo MF, Barra GB, Naves LA et al (2010) Selenoprotein-related disease in a young girl caused by nonsense mutations in the SBP2 gene. J Clin Endocrinol Metab 95:4066–4071

Barkoff MS, Kocherginsky M, Anselmo J et al (2010) Autoimmunity in patients with resistance to thyroid hormone. J Clin Endocrinol Metab 95:3189–3193

Beck-Peccoz P, Chatterjee VKK (1994) The variable clinical phenotype in thyroid hormone resistance syndrome. Thyroid 4:225–231

Beck-Peccoz P, Persani L (2010) TSH-producing adenomas. In: Jameson LJ, DeGroot LJ (eds) Endocrinology, adult and pediatric, 6th edn. Saunders Elsevier, Philadelphia, PA, pp 324–332

Beck-Peccoz P, Piscitelli G, Cattaneo MG et al (1983) Successful treatment of hyperthyroidism due to nonneoplastic pituitary TSH hypersecretion with 3,5,3′-triiodothyroacetic acid (TRIAC). J Endocrinol Invest 6:217–223

Beck-Peccoz P, Persani L, Calebiro D et al (2006) Syndromes of hormone resistance in the hypothalamic-pituitary-thyroid axis. Best Pract Res Clin Endocrinol Metab 20:529–546

Behr M, Ramsden DB, Loos U (1997) Deoxyribonucleic acid binding and transcriptional silencing by a truncated c-erbA beta 1 thyroid hormone receptor identified in a severely retarded patient with resistance to thyroid hormone. J Clin Endocrinol Metab 82:1081–1087

Bernal J (2007) Thyroid hormone receptors in brain development and function. Nat Clin Pract Endocrinol Metab 3:249–259

Bochukova E, Schoenmakers N, Agostini M et al (2012) A mutation in the thyroid hormone receptor alpha gene. N Engl J Med 366:243–249

Bradley DJ, Towle HC, Young WS III (1992) Spatial and temporal expression of α- and β-thyroid hormone receptor mRNAs, including the β2-subtype, in the developing mammalian nervous system. J Neurosci 12:2288–2302

Brucker-Davis F, Skarulis MC, Grace MB et al (1995) Genetic and clinical features of 42 kindreds with resistance to thyroid hormone. The National Institutes of Health Prospective Study. Ann Intern Med 123:572–583

Brucker-Davis F, Skarulis MC, Pikus A et al (1996) Prevalence and mechanisms of hearing loss in patients with resistance to thyroid hormone. J Clin Endocrinol Metab 81:2768–2772

Cheng SY, Leonard JL, Davis PJ (2010) Molecular aspects of thyroid hormone actions. Endocr Rev 31:139–170

Desouza LA, Ladiwala U, Daniel SM et al (2005) Thyroid hormone regulates hippocampal neurogenesis in the adult rat brain. Mol Cell Neurosci 29:414–426

Di Cosmo C, McLellan N, Liao XH et al (2009) Clinical and molecular characterization of a novel selenocysteine insertion sequence-binding protein 2 (SBP2) gene mutation (R128X). J Clin Endocrinol Metab 94:4003–4009

Dumitrescu AM, Refetoff S (2012) The syndromes of reduced sensitivity to thyroid hormone. Biochim Biophys Acta 1830:3987–4003

Dumitrescu AM, Liao XH, Abdullah MS et al (2005) Mutations in SECISBP2 result in abnormal thyroid hormone metabolism. Nat Genet 37:1247–1252

Fauquier T, Chatonnet F, Picou F et al (2014) Purkinje cells and Bergmann glia are primary targets of the TRα1 thyroid hormone receptor during mouse cerebellum postnatal development. Development 141:166–175

Ferrara AM, Onigata K, Ercan O et al (2012) Homozygous thyroid hormone receptor β-gene mutations in resistance to thyroid hormone: three new cases and review of the literature. J Clin Endocrinol Metab 97:1328–1336

Frank-Raue K, Lorenz A, Haag C et al (2004) Severe form of thyroid hormone resistance in a patient with homozygous/hemizygous mutation of T3 receptor gene. Eur J Endocrinol 150:819–823

Gurgel MH, Montenegro Junior RM, Magalhaes RA et al (2008) E449X mutation in the thyroid hormone receptor beta associated with autoimmune thyroid disease and severe neuropsychomotor involvement. Arq Bras Endocrinol Metabol 52:1205–1210

Gurnell M, Visser TJ, Beck-Peccoz P et al (2010) Resistance to thyroid hormone. In: Jameson LJ, DeGroot LJ (eds) Endocrinology, adult and pediatric, 6th edn. Saunders Elsevier, Philadelphia, PA, pp 1745–1759

Hamajima T, Mushimoto Y, Kobayashi H et al (2012) Novel compound heterozygous mutations in the SBP2 gene: characteristic clinical manifestations and the implications of GH and triiodothyronine in longitudinal bone growth and maturation. Eur J Endocrinol 66:757–764

Hammes A, Andreassen TK, Spoelgen R et al (2005) Role of endocytosis in cellular uptake of sex steroids. Cell 122:751–762

Hauser P, Zametkin AJ, Martinez P et al (1993) Attention deficit-hyperactivity disorder in people with generalized resistance to thyroid hormone. N Engl J Med 328:997–1001

Heuer H, Mason CA (2003) Thyroid hormone induces cerebellar Purkinje cell dendritic development via the thyroid hormone receptor alpha1. J Neurosci 23:10604–10612

Hodin RA, Lazar MA, Wintman BI et al (1989) Identification of a thyroid hormone receptor that is pituitary-specific. Science 244:76–79

Hu X, Lazar MA (2000) Transcriptional repression by nuclear hormone receptors. Trends Endocrinol Metab 11:6–10

Ito M, Roeder RG (2001) The TRAP/SMCC/mediator complex and thyroid hormone receptor function. Trends Endocrinol Metab 12:127–134

Kahaly GJ, Matthews CH, Mohr-Kahaly S et al (2002) Cardiac involvement in thyroid hormone resistance. J Clin Endocrinol Metab 87:204–212

Kapoor R, van Hogerlinden M, Wallis K et al (2010) Unliganded thyroid hormone receptor alpha1 impairs adult hippocampal neurogenesis. FASEB J 24:4793–4805

Kapoor R, Ghosh H, Nordstrom K et al (2011) Loss of thyroid hormone receptor β is associated with increased progenitor proliferation and NeuroD positive cell number in the adult hippocampus. Neurosci Lett 487:199–203

Lechan RM, Qi Y, Jackson IM et al (1994) Identification of thyroid hormone receptor isoforms in thyrotropin-releasing hormone neurons of the hypothalamic paraventricular nucleus. Endocrinology 135:92–100

Leonard CM, Martinez P, Weintraub BD et al (1995) Magnetic resonance imaging of cerebral anomalies in subjects with resistance to thyroid hormone. Am J Med Genet 60:238–243

Matochik JA, Zametkin AJ, Cohen RM et al (1996) Abnormalities in sustained attention and anterior cingulate metabolism in subjects with resistance to thyroid hormone. Brain Res 723:23–28

Mayerl S, Visser TJ, Darras VM et al (2012) Impact of Oatp1c1 deficiency on thyroid hormone metabolism and action in the mouse brain. Endocrinology 153:1528–1537

Mitchell CS, Savage DB, Dufour S et al (2010) Resistance to thyroid hormone is associated with raised energy expenditure, muscle mitochondrial uncoupling, and hyperphagia. J Clin Invest 120:1345–1354

Moran C, Schoenmakers N, Agostini M et al (2013) An adult female with resistance to thyroid hormone mediated by defective thyroid hormone receptor α. J Clin Endocrinol Metab 98:4254–4261

Neumann S, Eliseeva E, McCoy JG et al (2011) A new small-molecule antagonist inhibits Graves' disease antibody activation of the TSH receptor. J Clin Endocrinol Metab 96:548–554

O'Shea PJ, Bassett JH, Sriskantharajah S et al (2005) Contrasting skeletal phenotypes in mice with an identical mutation targeted to thyroid hormone receptor alpha1 or beta. Mol Endocrinol 19:3045–3059

O'Shea PJ, Bassett JH, Cheng SY et al (2006) Characterization of skeletal phenotypes of TRalpha1 and TRbeta mutant mice: implications for tissue thyroid status and T3 target gene expression. Nucl Recept Signal 4:e011

Ono S, Schwartz ID, Mueller OT et al (1991) Homozygosity for a dominant negative thyroid hormone receptor gene responsible for generalized resistance to thyroid hormone. J Clin Endocrinol Metab 73:990–994

Owen PJ, Chatterjee VK, John R et al (2009) Augmentation index in resistance to thyroid hormone (RTH). Clin Endocrinol 70:650–654

Phillips SA, Rotman-Pikielny P, Lazar J et al (2001) Extreme thyroid hormone resistance in a patient with a novel truncated TR mutant. J Clin Endocrinol Metab 86:5142–5147

Pulcrano M, Palmieri EA, Mannavola D et al (2009) Impact of resistance to thyroid hormone on the cardiovascular system in adults. J Clin Endocrinol Metab 94:2812–2816

Refetoff S, Dumitrescu AM (2007) Syndromes of reduced sensitivity to thyroid hormone: genetic defects in hormone receptors, cell transporters and deiodination. Best Pract Res Clin Endocrinol Metab 21:277–305

Refetoff S, Dumitrescu AM (2010) Resistance to thyroid hormone. In: Thyroid disease manager. http://www.thyroidmanager.org

Refetoff S, DeWind LT, DeGroot LJ (1967) Familial syndrome combining deaf-mutism, stippled epiphyses, goiter and abnormally high PBI: possible target organ refractoriness to thyroid hormone. J Clin Endocrinol Metab 27:279–294

Refetoff S, Weiss RE, Usala SJ (1993) The syndromes of resistance to thyroid hormone. Endocr Rev 14:348–399

Reutrakul S, Sadow PM, Pannain S et al (2000) Search for abnormalities of nuclear corepressors, coactivators, and a coregulator in families with resistance to thyroid hormone without mutations in thyroid hormone receptor beta or alpha genes. J Clin Endocrinol Metab 85:3609–3617

Schoenmakers E, Agostini M, Mitchell C et al (2010) Mutations in the selenocysteine insertion sequence-binding protein 2 gene lead to a multisystem selenoprotein deficiency disorder in humans. J Clin Invest 120:4220–4235

Schoenmakers N, Moran C, Peeters RP et al (2013) Resistance to thyroid hormone mediated by defective thyroid hormone receptor alpha. Biochim Biophys Acta 1830:4004–4008

Schomburg L, Dumitrescu AM, Liao XH et al (2009) Selenium supplementation fails to correct the selenoprotein synthesis defect in subjects with SBP2 gene mutations. Thyroid 19:277–281

Schroeder AC, Privalsky ML (2014) Thyroid hormones, T3 and T4, in the brain. Front Endocrinol 5:40 (eCollection)

Sjouke B, Langslet G, Ceska R et al (2014) Eprotirome in patients with familial hypercholesterol-aemia (the AKKA trial): a randomised, double-blind, placebo-controlled phase 3 study. Lancet Diabetes Endocrinol 2(6):455–463

Stein MA, Weiss RE, Refetoff S (1995) Neurocognitive characteristics of individuals with resistance to thyroid hormone: comparisons with individuals with attention-deficit hyperactivity disorder. J Dev Behav Pediatr 16:406–411

Takeda K, Sakurai A, DeGroot LJ et al (1992) Recessive inheritance of thyroid hormone resistance caused by complete deletion of the protein-coding region of the thyroid hormone receptor-beta gene. J Clin Endocrinol Metab 74:49–55

Torchia J, Glass C, Rosenfeld MG (1998) Co-activators and co-repressors in the integration of transcriptional responses. Curr Opin Cell Biol 10:373–383

van Mullem A, van Heerebeek R, Chrysis D et al (2012) Clinical phenotype and mutant TRα1. N Engl J Med 366:1451–1453

Visser WE, Friesema EC, Visser TJ (2011) Minireview: thyroid hormone transporters: the knowns and the unknowns. Mol Endocrinol 25:1–14

Weiss RE, Stein MA, Refetoff S (1997) Behavioral effects of liothyronine (L-T3) in children with attention deficit hyperactivity disorder in the presence and absence of resistance to thyroid hormone. Thyroid 7:389–393

Weiss AH, Kelly JP, Bisset D et al (2012) Reduced L- and M- and increased S-cone functions in an infant with thyroid hormone resistance due to mutations in the THRβ2 gene. Ophthalmic Genet 33:187–195

Index

CPSIA information can be obtained
at www.ICGtesting.com
Printed in the USA
LVOW05*2126151216
517432LV00002B/14/P